*Qianlong Garden Imperial Culture Series II*

# ARTISTIC WOOD DECORATION

Chief Editor

**Ancient Architecture Department of the Palace Museum**

Editor

*China Architectural Heritage*

# 《乾隆花园皇家文化系列·二——木艺奢华》编委会

主编单位： 故宫博物院古建部
承编单位： 《中国建筑文化遗产》编辑部

主　任： 单霁翔
顾　问： 晋宏逵　曹静楼　苑洪琪　刘　畅　吴子兴
委　员（按姓氏笔画排序）：　马国馨　王子林　王时伟　付清远　冯乃恩　刘若梅
　　　　　　　　　　　　　　刘　畅　李　越　张淑娴　苑洪琪　金　磊　高　志
　　　　　　　　　　　　　　晋宏逵　聂崇正　曹静楼　韩振平

丛书策划： 金　磊

主　编： 王时伟
副主编： 刘若梅　关　毅
编　辑： 文　明　宋永吉　赵丛山　洪　良
摄　影： 陈　鹤　田明杰　于跃超
装　帧： 计　什
翻　译： 宋悦兰　王丹青

*Qianlong Garden Imperial Culture Series II*
**Artistic Wood Decoration** Editorial Committee

| | |
|---|---|
| Chief Editor | Ancient Architecture Department of the Palace Museum |
| Editor | *China Architectural Heritage* Magazine |
| | |
| Director | Shan Jixiang |
| Consultants | Jin Hongkui　　Cao Jinglou　　Yuan Hongqi　　Liu Chang　　Wu Zixing |
| Members | Ma Guoxin　　Wang Zilin　　Wang Shiwei　　Fu Qingyuan　　Feng Nai'en　　Liu Ruomei |
| (by Stroke | Liu Chang　　Li Yue　　Zhang Shuxian　　Yuan Hongqi　　Jin Lei　　Gao Zhi |
| Sequence of Surnames) | Jin Hongkui　　Nie Chongzheng　　Cao Jinglou　　Han Zhenping |
| Series Planner | Jin Lei |
| | |
| Chief Editor | Wang Shiwei |
| Associate Editor | Liu Ruomei　　Guan Yi |
| Editor | Wen Ming　　Song Yongji　　Zhao Congshan　　Hong Liang |
| Photographers | Chen He　　Tian Mingjie　　Yu Yuechao |
| Designer | Ji Shi |
| Translators | Song Yuelan　　Wang Danqing |

# "乾隆花园皇家文化系列"总序

在中国古代建筑中，装修最精细的当属清代，而整个清代装修最精细的又是康乾盛世中的乾隆时期。可惜二百多年风雨过后，几经洗礼，具有"万园之园"之称的圆明园只存断壁，与宁寿宫势成犄角的敬胜斋毁于纵火，而西太后钟情的颐和园又因一再修葺而难觅盛世当年的风貌。如今人们能够看到的乾隆时期装修最精细的屋宇，仅存倦勤斋。它以其奢华和鲜明的个性风格遗世独立，更在整个皇宫980座、8704间房屋中别树一帜，成为探究盛清时代装潢艺术和乾隆内心世界的典范。尽管过去曾出版过《倦勤斋研究与保护》等专著，但今天面世的乾隆花园皇家文化系列不仅是一个传播故宫建筑的文化读本，更是从新层面开启人们探秘故宫的"大门"。

倦勤斋仿建福宫花园敬胜斋而建，斋名取"耄期倦于勤"之意。面阔九间，硬山卷棚顶，前有游廊与符望阁相连，形成东五间和西四间的格局。东五间内以木装修隔成凹字形转角仙楼。门后入口处留出开敞空间，形如广厅，其余则为上下两层仙楼。楼下正中设宝座，其余各间也设床榻。楼上各间相对独立，隔为封闭空间，以回廊相互连接。西四间相对开敞，西侧设一方亭式戏台。乾隆皇帝曾命南府（升平署前身）太监在此演唱岔曲。岔曲又称辇下小唱，以八角鼓、三弦伴奏，内容大多为歌功颂德、粉饰太平之作。戏台两侧设竹篱，为木雕髹漆工艺。西四间东侧与仙楼相接，楼上楼下西一侧设有宝座床，为观戏听唱之处。

倦勤斋仙楼木装修多为紫檀、花梨等名贵材料，装饰工艺包括镶嵌、竹黄、双面绣等多种，形成高贵中不失典雅的总体风格。双面绣也是清代出现的刺绣工艺新品种，其特点是绣面正反如一，可供两面观赏。倦勤斋内大量使用隔扇空间，其隔心部位使用双面绣装饰，图案秀雅，端庄中显出高贵，不同凡响。仙楼所用镶嵌工艺也别具特色，以竹丝、紫檀丝拼成万字锦地，嵌以碧玉，亦显典雅大方。特别要说明的是西四间最重要的装饰是覆盖天花和墙壁的通景画。北墙上绘有斑竹篱墙，与南侧木雕髹漆篱墙形成真假与虚实的对比。中间绘一月洞门，门外庭院有两只悠闲的仙鹤，庭院一侧巍然耸立一座楼阁，宫墙外显现出远山蓝天。西墙上也绘有竹篱、远山和常青的松柏。上顶天花部位满绘藤萝架，枝叶繁茂，盛开着淡紫色的花朵。花朵依远近透视而不同，形成奇妙而逼真的立体感。

据专家考证，该通景画为清代宫廷画师意大利人郎世宁的中国弟子王幼学所绘，其中也有部分郎世宁的手笔。通景画为分块绘制，贴裱拼接为一体，与欧洲的全景画和天顶画有异曲同工之妙。画面之大，画艺之精，为国内仅存的孤品。

倦勤斋的内装修经过了特别精心的设计制作。乾隆花园的修建年代正值清代政治稳定、经济繁荣的阶段。皇家的建筑不惜人力物力。如内务府大臣英廉、福隆安为工程处总理，从设计、绘图到烫制小样，全部经过御览、钦定、"照样准做"等一系列过程。凡事以满足乾隆皇帝的欲望和要求为标准。此时的乾隆皇帝醉心于苏杭的秀丽风光，倦勤斋内的细部设计均由宫内量准尺寸"定身打造"，发样交江南地方督办，然后运至北京组装。有些江南物产不适应北京的干燥气候，如"竹制"装修在北方常常会离缝走样、脱落爆裂。于是在倦勤斋室内营造的自然园景中，凡需要斑竹原材的部分全部采用木雕及髹漆斑竹花纹的方法。工匠们不厌其烦地逐一绘出了满庭"斑竹"，以求传达江南风情，博得龙颜欢悦，也恰为此，在今人眼中，它们才美轮美奂。

乾隆花园（宁寿宫花园）是乾隆皇帝直接设计并指挥建造的。他深厚的文化修养使他把诗情与画意、皇家园林与私家园林、南方与北方工艺、东方与西方装饰手法有机融汇在一起。为此故宫博物院古建部在深入研究解读乾隆时期皇家文化深刻内涵的基础上，与《中国建筑文化遗产》杂志社、天津大学出版社合作策划，决定陆续出版"乾隆花园造园艺术研究"、"乾隆花园室内装饰艺术研究"、"乾隆花园室内家具研究"、"乾隆花园室内装修图集"、"乾隆花园建筑彩画研究"等系列图书。本系列读物，并非学术著作，它旨在通过经典的文字与精彩的图片、版图向读者展示乾隆皇帝的精神世界和皇家建筑园林艺术、室内装饰艺术的精髓等内容。相信，读者定会从中发现故宫建筑独特的美，更深刻地赏析到中华文化的博大精深。

2012年10月正值故宫博物院建院八十七周年，这套围绕乾隆花园皇家文化所展开的造园艺术、室内装饰、家具研究、彩画分析丛书得以创作与传播，尤其希望它们能在向海外传播故宫世界文化遗产方面发挥独特作用。是为序。

故宫博物院院长

2012年9月8日

# General Preface of *Qianlong Garden Imperial Culture Series*

Qing-style constructions in all Chinese ancient architecture are evidently the finest, their decorations are found the most exquisite in material and style during Qianlong period of Kangxi and Qianlong Golden Age. However, after over two hundred years since the accomplishment of the buildings in Qing Dynasty, Yuanmingyuan Imperial Garden has only ruins remained, Jingshengzhai within Jianfugong Garden, which is situated on the opposite corner to Ningshou Palace of the Forbidden City, was damaged in a fire, the original style and features of the Summer Palace that the Empress Dowager loved so much has gone after repairs. Juanqinzhai (Studio of Exhaustion from Diligent Service) is so far the only building which retains the most exquisite decorations of Qianlong period. Juanqinzhai is so unique in its characteristic that it has been a sample for specialists to study the decoration art of Qing Dynasty and the Qianlong Emperor's inner world because the interior decorations are luxurious and peculiar among the 980 Imperial Building Complexes. Despite of the publication of *Juanqinzhai Research and Conservation* and other books, the *Qianlong Garden Imperial Culture Series* are not only books to distribute Imperial Culture, but also to open the mysterious "door" to the public.

Juanqinzhai was built imitating Jingshengzhai within Jianfugong Garden (locates in the northwest of the Forbidden City), named "Studio of Exhaustion from Diligent Service". Situated in the most northeast of the Forbidden City, Juanqinzhai is a building of nine bays, divided into two sections, five bays in the east and four bays in the west, with saddle and rolled pitched roof, a corridor in front leading to Fuwangge to its south. The wood partitions form concaved-attic in the east section. The central bay behind the door is an open space, like a hall, and the other bays are divided into upper and lower storeys. A throne facing to the door is located in the central bay on the lower storey, and couches are occupied in the rest. The space on the upper storey is separated into enclosed rooms, connected with corridor. The west section is open; a square pavilion theatre stage stands on the west side. The Qianlong Emperor ordered eunuchs to perform *Chaqu*, short performances accompanied with octagonal drum and three-string instrument. The content is mainly praise of the wise emperors and peaceful life. Faux bamboo fences stand at both sides of the theatre. A seating platform, facing the stage, is placed at the east side of the west section, from which the Emperor would watch performances either on the upper or lower storey.

The screen-partitions of concaved-attic in the east section are mainly decorated with precious wood such as red sandalwood, padauk, mahogany, etc. The decoration technique includes inlays, inner bamboo skin veneer and double-sided embroidery and so on, presenting a style of grandeur but elegance. Double-sided embroidery as a new technique appeared in Qing Dynasty, demands the same evenness and smoothness on both sides of the silk. Many pieces of double-sided embroidery are used in screen-partition panels, which provide a unique sense of opacity and translucency. The delicate, colorful stitches depict auspicious flowers, leaves on branches, fungus and geometric patterns reminiscent of ancient bronze patterning. The attic decorations are quite distinctive, bamboo thread marquetry, *zitan* strips swastika pattern and jade carving are inlayed to be classical and elegant. What should be specially mentioned is the huge panoramic mural painting overall mounted on the walls and ceiling in the west section. The speckled-bamboo fence on the north wall mural matches the south faux bamboo real fence, a moon-shaped gate in the middle of the north wall mural painting echoes the real one in the south fence, two cranes play leisurely out of the gate, a grand pavilion stands majestically in the courtyard, mountains and blue sky are seen farther into the distance. The west wall mural painting presents bamboo fence, mountains and evergreen trees beyond. While looked up, the entire ceiling is covered by a painting depicting a speckled-bamboo trellis, from which hang clusters of pink and violet wisteria blossoms surrounded by green leaves. The flower clusters gradually lean away to appear in three dimensions while looked up towards the ceiling.

Based on historians' research, the panoramic mural paintings within Juanqinzhai were painted by the Imperial artist Wang Youxue, who had learned foreshortening technique from an Italian missionary, Giuseppe Castiglione, known in Chinese as Lang Shining, he came to China during Qing Dynasty. The massive mural painting is painted in pieces and mounted together as a whole, quite similar to the Trompe-l'œil panoramic paintings in Europe. The mural painting in Juanqinzhai is so large that it is the only one extant in China.

Juanqinzhai interior decorations were meticulously designed and manufactured. Qianlong Garden was constructed while the country was in its zenith period, politics stable and economy prosperous. Therefore, the Imperial Court would not hesitate to invest any human power and materials to meet its requirements. The Imperial Household Ministers such as Ying Lian and Fu Long'an were then the General Manager of the Construction Project. They supervised the whole process of the construction from design, drawing to sample making after the Emperor's reading, approval and decision. The Qianlong Emperor was so obsessed with Jiangnan areas like Suzhou and Hangzhou with beautiful scenery that he ordered the Jiangnan local officials to make Juanqinzhai interior decorations particularly according to the design and dimensions given by the Imperial Court, then, ship to Beijing for installation afterwards. However, the bamboo-material decorations became loose, detached, cracked, even deformed due to the dry weather in the north. In order to create beautiful Southern scene inside Juanqinzhai to satisfy the Emperor, all bamboo-material fences and windows were made with *nanmu* wood and then painted into faux speckled-bamboo. Therefore, there is no doubt people today find it so striking.

Qianlong Garden (also called Ningshougong Garden) was constructed under the Qianlong Emperor's direct design and instruction. He ingeniously merged poetry with painting, the Imperial garden with private garden, Southern technique with Northern technique, Eastern decorations with Western decorations. Experts from the Ancient Architecture Department of the Palace Museum did thorough research and study based on the archival documentation of the Imperial Culture during Qianlong period and begin to publish a series of books, cooperating with *China Architectural Heritage* Magazine and Tianjin University Press. These books will involve *A Study of Qianlong Garden Gardening Art, A Study of Qianlong Garden Interior Decorations Art, A Study of Qianlong Garden Furniture, A Collection of Drawings and Pictures of Qianlong Garden Interior Decorations, and A Study of Qianlong Garden Architectural Color Painting*, etc. This series of publications will demonstrate to readers both texts and pictures of the Qianlong Emperor's inner world, the essence of Imperial architecture and gardening, interior decorations. One should believe, readers would surely learn about the unique beauty of Imperial Architectural Complex, appreciate such profound Chinese Culture to a deeper extent.

October of 2012 is the 87th anniversary of the establishment of the Palace Museum, the publication and distribution of this series of books at this moment are to play a distinctive role in dissemination of the Palace Museum as a World Cultural Heritage, so the preface is written.

*Director of the Palace Museum* **Shan Jixiang**
2012-9-8

# 前　言

以木材制成器物，古已有之。早在距今约7000年的浙江河姆渡文化遗址的考古发掘中，发现了雕花木桨、剑鞘、圆雕、木鱼、鱼形器柄等器物。3000多年前的殷商时期出现了中国建筑木雕，木雕与木架结构紧密结合，进入了"大木作"的初始阶段。

汉末以前，由于以跽坐为主的起居方式，席与床（榻）是主要的室内陈设。汉代地位较高的人在床上加的帐、几、案比较低矮，屏风多用于床上。南北朝已有高型坐具，唐朝出现了高型桌、椅和落地屏风。经五代到宋定型，唐宋时期的木雕工艺在室内装饰领域开创了新天地，家具随着人们起居习惯的演变而变化。唐宋以后，垂足而坐的习惯形成。随着椅的产生，桌、几、案加高，上面的摆件也随之增多，大木作与小木作在制作工艺上各司其职。大木作统领的格局产生变化，木作行业形成了进一步的细分。

明清宫廷家具及木雕装饰的形成和发展得益于社会经济的发展，是在民间家具木雕技艺空前发达的基础上发展起来的。它来自民间，又高于民间。由于工匠技艺精湛，用材精良，因此宫廷家具木雕装饰代表了中国传统家具木雕装饰的最高水平。

近年来，故宫博物院开展了"乾隆花园保护项目"的硬木装修、家具的修复工作。遵循文物修复原则，研究总结传统修复工艺材料，使大量硬木装修、家具得到良好的保护修复。

《木艺奢华》一书展现了乾隆时期以紫檀木等为依托体，各种工艺、材料融为一体的家具、装修艺术品，这些艺术品充分反映了乾隆时期木雕工艺、装饰工艺的水平，有很高的艺术观赏性。

王时伟

2013年11月8日

# Foreword

There has been quite a long history since the existence of wood objects. Wooden objects such as carved wooden paddle, wooden sheath, sculpture, wooden fish, fish-shaped handle were discovered in the archaeological excavation of Hemudu Site Cultural Ruins, which existed about 7000 years ago. About 3000 years ago, during the Shang Dynasty, woodcarving on architectural structure emerged. Since then, woodcarving and wooden architectural structure closely bonded together, marking the beginning of "carpentry work".

Due to the life style of kneeling in Han Dynasty, most of the interior furniture are mats and beds. Beds used by people of high social status would be decorated with curtains and tables which were comparatively lower .The screens were mostly placed on bed. In Northern and Southern Dynasties, seats in taller form came into being. High tables, chairs and standing screens were not created until Tang Dynasty. In Tang and Song Dynasties, woodcarving on interior decoration and furniture evolved with the development of living habits of human being. From Tang and Song Dynasties, living habit of sitting was gradually developed. With the emergence of tall chairs, tables become taller and the decorative objects displayed on tables were increased. Techniques of carpentry work and joinery work were developed within their own fields. The domination of architectural woodworking changed and woodworking industry was divided into more branches.

Formation and development of woodcarving decoration and imperial furniture of Ming and Qing Dynasties were benefited from the development of social economy, based on the excellent carving technique of folk furniture. Imperial carving technique originated from folk technique, but exceeded it. Owing to the exquisiteness of the craftsmanship and the fineness of the material, woodcarving decoration on the imperial furniture leads the way of the most advanced level of traditional furniture carving.

In recent years, the Palace Museum of China initiated the conservation on wooden interior decorations and furniture of Qianlong Garden Conservation Project. The conservation follows the repair principle and carried out a great deal of researches on traditional materials, so that massive interior decorations and furniture could be well conserved.

Art of Wood exhibits furniture and decorative arts made of red sandalwood and other categories of wood, combining with a wide range of materials and techniques, which highly reflects the advancement of woodcarving and decoration during Qianlong period.

Wang Shiwei
2013-11-8

# 目录

"乾隆花园皇家文化系列"总序 · 单霁翔
前言 · 王时伟

## 第一部分 · 综述
清乾隆时期木雕工艺 ········································································ 016

## 第二部分 · 图版
竹木雕刻 ························································································ 044
竹丝镶嵌 ························································································ 152
画珐琅镶嵌 ···················································································· 158
錾铜镶嵌 ························································································ 162
螺钿镶嵌 ························································································ 166
玉石、宝石、大理石镶嵌 ······························································ 176
雕漆工艺 ························································································ 246
多种材质及其工艺 ········································································· 250

## 第三部分 · 附录
硬木家具装饰修缮技艺 ··································································· 274

# Contents

**General Preface of Qianlong Garden Imperial Culture Series** · Shan Jixiang
**Foreword** · Wang Shiwei

## Chapter I  General Introduction
The Woodcarving Handicraft of Qianlong Period in Qing Dynasty ........ 016

## Chapter II Drawings
Bamboo and Wood Carving ........ 044
Bamboo Filament Marquetry ........ 152
Painted Enamel Inlay ........ 158
Chiseled Copper Inlay ........ 162
Mother-of-pearl Inlay ........ 166
Jade, Precious Stone and Marble Inlay ........ 176
Lacquer Carving Technique ........ 246
Various Materials and Techniques ........ 250

## Chapter III Appendix
Restoration Technique on Hardwood Furniture ........ 274

崇敬殿内景陈设
Interior Scene of Hall of Adoration (Chongjingdian)

# Chapter I General Introduction

第一部分 · 综述

# The Woodcarving Handicraft of Qianlong Period in Qing Dynasty

# 清乾隆时期木雕工艺

清乾隆时期木雕工艺融合明代工艺木雕的形制结构，在造型上强调稳定、厚重；装饰题材上采用寓意丰富的吉祥瑞庆内容，展现出人们对生活的美好愿望和幸福追求；制作手段上集纳锼镂雕刻、剔錾榫卯、嵌镶描绘等高超技艺。选料精当，设计精巧，制作精良，榫卯牢固，纹饰精美，舒适耐用，代表着创意设计、工艺技术的最高水平。

小木作，大致可分为三大类：一是小器作的木座、陈设摆件等实用工艺美术品；二是家具；三是室内外装饰。

## 小木作与大木作

宋代《营造法式》已把门、窗、隔扇、藻井、天花板、屏风、楼梯及栏杆等列入小木作范围，并对雕工的技艺进行了具体规范，将"雕作"按雕刻形式分为四类（即混作、雕插写生华、起突卷叶华、剔地洼叶华）。按雕刻技术可分为五种基本形式（即混雕、线雕、隐雕、剔雕、透雕）。所以要突出小木作的特点，离不开木雕工序，而大木作专攻建筑木结构而无须雕刻工的配合。

对页图：重华宫内景陈设
Opposite Page: Interior Scene of Palace of Double Brilliance (Chonghuagong)

The furniture woodcarving handicraft of Qing Dynasty in Qianlong Period combines the characteristics of Ming style in form and structure, and emphasizes the stability and heaviness. The decorations demonstrate auspiciousness and happiness indicating hope and expectation; the technique composes of open work carving, chiseling, jointing, inlaying and lacquer painting. The furniture is planned for exquisite materials, elaborate design, delicate manufacture, stable joints, auspicious patterns, comfort and durability. The design and technique represent the highest level at that time.

Joinery work is generally divided into three types. The first is the wooden bases for small objects, the second is furniture, and the third is interior and exterior decorations.

## Joinery work and carpentry work

Woodwork such as doors, windows, partitions, caissons, ceilings, screens, stairs, and railings are defined as joinery work in the book of *Ying Zao Fa shi*( an offical scientific monograph on architecture and construction in Song Dynasty). Carving craft is specified into four types according to carving forms and five types according to techniques. So joinery work carving won't be fine without carving. And carving has to be carried out in accordance with procedure, and the carpentry work mainly specializes in architectural wooden structure, which

综述・清乾隆时期木雕工艺

第一部分

清代《工程做法》把门窗、隔扇等称为装修工程，并分为外檐装修和内檐装修，而木雕则侧重室内木雕装饰。清康乾盛世，故宫的室内木雕装饰堪称举世无双的艺术佳作，是现代装饰中经常借鉴的典范。

## 小木作的工艺范畴

小木作的工艺范畴分为花活、木活、大件、小件。"花活"是有雕刻的木制品，以雕刻为主；"木活"是无雕刻或只有少量雕刻的素活，以木工为主；"大件"是家具和室内装饰木雕；"小件"是把玩件和陈设木雕及木座类的小器作产品。

**1. 小器作包括木座、几案陈设**

（1）木座

素座（圆形、方形、随形）、花座（起地花座、嵌银丝座、多台花座）天然座（山座、水座、云座、把莲座、荷叶座、莲花座、树根座）等。

（2）架子

盘架、合页架、十字架、提梁卣和链瓶架等。

（3）几案陈设

笔筒、笔架、镇尺、臂搁，托盘、承盘、棋盘、棋盒、砚盒、书画盒、首饰盒、食盒，印匣、拜匣、皮箱、药箱、文具箱、案屏、镜架、桌灯、如意以及种类繁多的圆雕动植物木摆件等。几案陈设基本以圆、浮雕摆件和实用木制工艺品为主。

**2. 家具和室内外木雕装饰**

（1）家具

家具包括凳椅、几案、橱柜、床榻、架和屏风六大类。

（2）室内外木雕装饰

室内外木雕装饰有藻井、天花、隔扇、窗扇、落地罩、几腿罩、天穹罩、垂花门、太师

对页图：重华宫内景陈设
Opposite Page: Interior Scene of Palace of Double Brilliance (Chonghuagong)

does not need the cooperation of woodcarving.
In Qing Dynasty, the book *Engineering Practice* defined the doors, windows, partitions and screens as decoration engineering, including exterior and interior decorations; woodcarving is primarily manufactured for interior decoration. Then, woodcarving decorations found in the Forbidden City of Kangxi and Qianlong Periods are regarded as unique fine and exquisite art pieces, examples for modern decoration.

## Category of handicraft of joinery work

### Tiny objects are set on wood bases, and objects are usually displayed on tables

i) Bases
Bases without carving shaped into round, square, oval, and the bases with carving, silver, thread inlay and superimposed layers; bases carved into natural objects such as mountains, rivers, clouds, lotuses, lotus leaves, lotus flowers and tree roots.

ii) Shelves
Plate racks, hinge brackets, cross frames, wine pot racks and chain bottle frames.

iii) Objects on tables
Brush pots, penholders, paper-weight, arm-sets, trays, chess boards, chess boxes, cases, screens, lamps, ruyi( an S-shaped ornament object ), animal and plant shape objects. They are usually carved into round shape, relief sculptures and pratical art works.

### Furniture, exterior and interior woodcarving decorations

i) Furniture
Furnitures include chairs, stools, tables, cabinets, beds, shelves and screens.

ii) Exterior and interior woodcarving decorations
Exterior and interior woodcarving decorations are caissons, ceilings, screens, windows, partitions, doors, handrails, columns, and so on.

Large groups of artisans engaged in creative design

壁、墙饰、柱饰和栏杆等。

明代"御用监"、清代"造办处"集中了一大批参与宫廷艺术创意设计的高级文人，同时把已形成木雕艺术氛围的广州的"广作"、苏州的"苏作"艺人请进北京。艺人的技术与宫廷画师的设计相结合，把中国文人木雕艺术推向了高潮，形成了独具宫廷风格的木雕艺术流派——宫廷木雕，也称为"京作"。

（1）"京作"定义

在京都代表京城地域特点和宫廷风格的工艺作坊。

（2）"京作"特点

一是拥有珍贵的木材，二是高级文人参与设计，三是能工巧匠的聚集，四是不计成本的精工细作。金、玉、象牙、雕漆、珐琅等多种材料相互结合，产生了新的艺术形式，如在木门窗和家具上镶嵌玉、象牙、珐琅等。

（3）"京作"的作用

宫廷木雕代表着时代的繁荣和技艺的进步，是艺术的结晶和雕刻技法的体现。"京作"木雕和全国各流派木雕有着千丝万缕的密切联系，同时引导着中国广大地域的各流派的形成和发展。福州、潮州、苏州、东阳四大木雕都是在明、清两代进入发展高潮，而且都受到"京作"艺术的启发和引导。

## 宫廷木雕

宫廷木雕是指皇家组织生产的木制品。尽管工匠多数来自江南，但在北京的皇权制度下高度统合，有其独特的风格和艺术内涵，受宫廷、士风影响产生了特殊艺术，代表着皇家艺术的特色。

为满足宫廷木雕工艺品的大量需求，明、清两朝采用了以下三种办法。

### 1. 内造

明朝设"御用监"、清朝设"造办处"，在宫内直接监造。清康熙年间，"造办处"把一处木作

for imperial court from all directions of China were selected to the imperial court during Ming and Qing Dynasties. Meanwhile woodcarving artisans of Guangzhou-style from Guangzhou and Suzhou-style from Suzhou were invited to Beijing. Carving technique combined with imperial painters' design pushed woodcarving art into a higher level, and a unique woodcarving style, Beijing-style, came into being ,which represented the imperial level.

i) Beijing-style workshop
Beijing-style workshop presented the woodcarving characteristics of Beijing and Imperial Palace.

ii) Beijing-style's characteristics
Beijing-style's characteristics are well-known for precious materials, literates' participation into the designing, artisans who from all directions gathered together, extravagant and elaborate craftsmanship. Materials such as gold, jade, ivory, lacquer, cloisonne and so on produced a new art form. For example, jade, ivory,cloisonne were insetted in wooden doors,windows and furnitures.

iii)Beijing-style workshop function
Imperial woodcarving demonstrated prosperity of that period and progress of the technology. Beijing-style workshop was closely connected with the techniques and styles of the rest part of the country, but took the leadership as well. Woodcarving in Fuzhou, Chaozhou, Suzhou, and Dongyang developed and thrived during Ming and Qing Dynasties under the influence of Beijing-style workshop.

## Imperial woodcarving

Imperial woodcarving is the wooden articles made in Imperial Household Department by the craftsmen from Jiangnan region, and controlled and influenced by the imperial court requirements both on style and standard.Its unique style and artistic meaning represented the characters of royal art.

Three ways for manufacturing wooden products

设在紫禁城内西南角的武英殿北、慈宁宫南、隆宗门西，另一处设在圆明园的芰合香与紫碧山房一带。

**2. 委造**

委托制造，皇家画院出样式，传交各地方政府承做的官作家具。

**3. 贡造**

各地根据宫廷式样制作进贡给皇室的家具。委造与贡造虽然属于外造，但它与内造一样严格按规范制作，也属于宫廷木雕，同样代表着宫廷木雕的艺术风格。贡造制作的木雕艺术精品是对宫廷木雕的补充。

据宫廷档案记载，"造办处"的木作不单制作家具，也承制木制日用品，并负责为玉器、金银器、古董摆件配底座，还包括制作一些递送奏折的匣子，各种盒、架及木雕摆件。

木作还为贵重工艺品制作木质式样。如被冠以工艺与材质之王的特大玉雕《大禹治水》，就是先由木雕工根据玉石料原形，按照图案设计要求全部用木料雕成，作为烫样参照，然后再在玉石上施以雕刻。

## 创艺设计

根据《周官》与《左传》的记载，周王和诸侯都设有掌管营造的司空。此后，各个朝代都沿袭了这种工官制度。主管具体工作的官吏，《考工记》称匠人，唐朝则称大匠，主要从事设计和主持制作的工匠称都料匠。明朝还有少数工匠出身的工部首脑人物。清朝"造办处"各级职守，分为主管、领催（工长）、催头（工头）、工匠。

宫中木作的运作有一套严格的审批程序。以清代为例，首先由（如意馆内）宫廷画师拿出设计画样，有时要提供两三套设计方案，供对比选择。经皇帝阅览、修改，批"照件准做"后，才得以施工。"造办处"必须把设计、批件、施工、验收、修改过程记录在案。"造办处"的管理人员，如怡亲王允祥、海望，宫

were adopted to meet the large needs of the Imperial Household in Ming and Qing Dynasties.

### Imperial manufacture

The Imperial Household Department in Ming and Qing Dynasties supervised and produced furniture and decoration articles for imperial household supply. During Kangxi Period of Qing Dynasty, one of the imperial woodwork workshops was situated at the southwest part of the Forbidden City, the other in northwest part of Beijing.

### Appointed manufacture

The Imperial Household Department appointed local governments to produce articles according to the samples provided by the Imperial Painting workshop.

### Tribute articles

Local governments paid tribute to the Imperial Court. Tribute articles and appointed manufacture were made in accordance to the imperial requirements, so they represented the technique and art of imperial workshop.

Based on the imperial archival, the woodwork workshop of Imperial Household Department produced not only the furniture, but also some daily-used articles, bases for objects of jade, gold, old antiques, and boxes for memorial to the throne, frames for objects.

Woodwork workshop, in addition, provided wood samples for precious objects. For example, the large piece of jade carving work of *King Yu Combating the Flood* was manufactured after the wood sample was made on the basis of the original shape of jade.

### Creativity and Design

Books of *Zhou Guan* ( Classic on Zhou's System of Officials) and *Zuo Zhuan* ( The Commentary of Zuo to the Annals of Spring and Autumn) recorded that

廷艺术家刘源、唐英及意大利人郎世宁、法国人王至诚、沙如玉曾参与当时的木作式样设计。

清代宫廷家具制作中的一个现象是帝王的直接参与，尤其以乾隆帝最为积极。宁寿宫花园所有装修家具都是在他耐心、细致的过问下设计、选料、制作、修改后完成的。乾隆三十年档案载有如下史实："交紫檀木镶摆锡玻璃檀香花纹宝座一座，传旨将宝座上玻璃并檀香花纹俱拆下，另画样呈览。"

## 制作工匠

木作工匠多数来自广东、江苏，由地方选拔保举进京，经"造办处"试用合格后正式录用。没有过硬的本领和技术上的绝活，就无法通过层层技术考核。宫廷之内可说是群英荟萃，能进宫做工的一定是有实践经验的技术高手。这样的人才，大多数被安置在距紫禁城不远的皇城根一带，进京时地方政府和皇家还要发给工匠们安家费，并可以取得内务府抬籍，得到丰厚的工银。工匠每年有回乡扫墓假，工钱照发，报销路费。在封建社会中，除了宫廷工匠外，其他行业不可能有如此优厚的待遇。乾隆皇帝不占工匠便宜，有"料给值，工给价"的说法。按户部俸禄标准，木作工匠的工银高于知县的俸禄。在这样的条件下，无后顾之忧的工匠全身心地施展才艺，并不断提高自己的技艺水平。"造办处"根据质量讨赏，做得好先"拟赏"，呈皇帝审批。御批的赏赐大多高出"拟赏"，以示皇恩。宫廷木雕的用料质量和技术的高标准要求，也促使匠人间进行技术比拼，从而促进技艺不断提高。

宫廷木雕属高层文人木雕工艺与民间的文人木雕工艺不同。"造办处"管辖的"錾花作"、"珐琅作"、"镀金作"、"漆作"、"牙作"等可以相互提供配件，合作完成作品，具有民间文人木雕无法比拟的技术支撑。

对页图

上：漱芳斋内戏台
Above: Theatre Stage in Lodge of Fresh Fragrance (Shufangzhai)

下：翠云馆宝座陈设
Below: Interior Display of Throne in Cuiyunguan

King Zhou and princes set up department in charge of imperial manufacture, so did the following dynasties. The responsible officers are nominated as masters in book of *Record of Handicrafts Manufacture*. Masters in Tang Dynasty dealt with both design and manufacture; some masters in Ming Dynasty, became officers of the Board of Works eventually. Then, in Qing Dynasty, craftsmen in the Imperial Household Department were organized and nominated in different positions based on their talents and skills.

A piece of woodwork or furniture is completed by the imperial workshop under strict procedures, after series of approval. Taking Qing Dynasty for example, imperial painters provided a drawing sample, sometimes two, or even more for carving masters to compare and select. The work started when the emperor mended and finally approved. The Imperial Household Department had to record all the processes of sampling, mending, approving, checking and accepting. The managers recorded in archive participating in the wood workshop design were the Princes Wang Yunxiang and Hai Wang, Imperial artists Liu Yuan, Tang Ying, Italian priest Giuseppe Castiglione, French Jean Denis Attiret and Valentin Chalier.

It is interesting that the emperors would participate in the design of woodwork themselves during Qing Dynasty, especially the Qianlong Emperor. He took troubles to check the design, material, manufacturing, and revision. The imperial archive of Qing Dynasty recorded that Qianlong Emperor once ordered to redesign and present for his checking after removing the glass and sandalwood decoration.

## Manufacturing artisans

Most woodwork artisans came from Guangdong and Jiangsu based on local selection and recommendation, then they were employed by the Imperial Household Department after the probation. One had to pass tests one after another to prove that he was capable and skilled enough. Those who were eventually employed would settle down nearby outside of the Forbidden

## 装饰风格

以清代家具为例，其造型庄重，气势非凡，力求富丽华美，形式富于变化，品种日益增多，多种工艺结合，成为注重与室内空间相协调的实用观赏艺术品。

宫廷木雕多用珍稀材料制作，名家设计，高手雕刻而成，有严密的管理制度和严格的制作质量标准，因此制作成本高昂。

明、清时期，大批文化修养很高的文人参与了木雕的设计。文人工艺美术爱好者参与木雕设计，提高了木雕工艺的档次。他们把对书法艺术的理解和审美，融合在家具陈设的设计、装饰中，促进了木雕艺术的繁荣。由于文人亲自参与制作，在雕刻中融入自己的画风，并用雕刻的方法表现出画笔所不及之处，因此这种雕刻风格尽脱匠气，是有高雅品位的艺术品，称为文人木雕。宫廷木雕最高的形式就是文人木雕。

### 1.龙纹

龙是中国最大的神物，古人认为它是最神圣的祥瑞。在古文献中有很多这方面的记述，有关它们的传说亦不少，大多离奇神怪，众说不一。在殷人的卜辞里，有很多龙字，龙是殷人卜问的对象之一，是殷人崇拜的百神之一。龙纹原先是神武和力量的象征，后在封建社会被作为"帝德"和"天威"的标志，不准乱用。《说文解字》：龙"能幽能明，能细能巨，能短能长"，能"兴云雨，利万物"，"注雨以济苍生"。《礼记·礼运》："麟、凤、龟、龙，谓之四灵。"《本草纲目》："龙者鳞虫之长。王符言其形有九似：头似驼，角似鹿，眼似兔，耳似牛，项似蛇，腹似蜃，鳞似鲤，爪似鹰，掌似虎，是也。背有八十一鳞具九九阳数，口旁有须髯，颔下有明珠，喉下有逆鳞。头上有博山"

龙的图案在先秦以前，形象较质朴粗犷，

对页图：龙纹
Opposite Page: Dragon Pattern

City; the local government and the imperial court would provide financial relief to them. The artisans were offered pensions and holidays to visit their hometown once a year; they were paid well, even better than a county chief. Therefore, these artisans made a big effort to work with great enthusiasm; their potentials were played to the greatest extent. They competed with each other and improved their ability, skills, and techniques constantly.

The imperial woodcarving, which emphasized more on literate taste, is different from the folk carving. There are workshops for different sorts of techniques such as chisel carving, cloisonne, gold gilding, lacquer, ivory, and so on. These workshops usually cooperated together in order to complete a piece of work, whereas the folk carving can hardly catch up with it.

### Decorative styles

The Decorative styles on woodwork in Qing Dynasty are made firm and heavy in size, luxuriant and exquisite in style, changing and diverse in form. Decorations of interior echo with that of exterior neatly with the increase of the varieties of furnitures and the combination of varieties of techiques.

The imperial woodcarving articles are made of precious and rare materials, designed by well-known artists, carved by great masters, so how great the expense can be imagined.

In Ming and Qing Dynasties, many literatus were involved in the design of woodcarving. Their understanding and taste of calligraphy were integrated into the design and decoration, which promoted woodcarving art. Because of their participation, the woodcarving articles represent not only the technique of masters but also artists' aesthetic level. The objects produced for display, decoration, and daily supplies, in that case, they are artistic works rather than objects. These works are master pieces among the imperial woodcarving articles.

综述 · 清乾隆时期木雕工艺

第一部分

025

大部分没有肢爪，近似爬虫类动物。秦汉时的龙纹多呈兽形，肢爪齐全，似虎似马，常作行走状。发展至隋唐时，龙的嘴角和腿部特长，尾部似蛇。宋代龙纹和唐代龙纹基本上近似，下颚开始上翘。元代时，出现飘拂状的毛发，腿部有"露筋露骨"的纹饰。明代时，筋骨演变为在腿上全部拉线，头上毛发上冲，龙须外卷或内卷，并出现风车状五爪。清代时，龙头毛发横生，出现锯齿形腮，尾部有秋叶形装饰。明清家具上的龙纹与时代一脉相承，较多地见于宫廷中使用的家具，且大都规整威严，气势雄伟，以云龙纹饰最为普遍。

**2. 缠枝纹（卷草纹）**

缠枝纹为传统吉祥纹样，又名"万寿藤"，寓意吉庆。因其结构连绵不断，故又具"生生不息"之意，是将一种藤蔓卷草提炼变化而成，委婉多姿，富有动感，优美生动。缠枝纹约起源于汉代，盛行于南北朝和隋唐，宋元和明清时期应用也比较广泛。

缠枝纹以牡丹组成的称"缠枝牡丹"，以莲花、葡萄组成的称"缠枝莲"、"缠枝葡萄"，以人物和鸟兽组成的称"人物鸟兽缠枝纹"。明清时缠枝花纹严谨工整，丰富华丽，是明清家具常见的装饰纹样之一。其中最简单的形式常称为"卷草纹"，有的仅将草叶稍加弯曲变化，作对称式或连缀式，极易与家具的线脚呼应协调，故运用非常广泛。

**3. 松竹梅**

古人爱松竹梅，追求的是其象征的精神，这是三种在寒冬时节依旧保持顽强生命力的植物。但严格来说，梅花却并不是迎寒开花，在冬季也并不能像苍翠松竹那样让人感受到其生命力有多顽强。

梅花是在早春迎暖而开，倒是腊梅、茶梅

**Dragon pattern**

Dragon is the greatest deity for Chinese; the ancestors regarded it as the highest auspicious omen. Records and legends about it are found in many documents, and most are supernatural with miraculous power. Dragon is seen in divination as one of Gods for Yin people to worship. Dragon pattern was firstly taken as a symbol of spirit and power, later a symbol of emperor's prestige and authority, so no one dare to use it arbitrarily. *Analytical Dictionary of Characters* says: "Dragon is able to be visible and invisible, big and small, long and short; bring cloud and rain, prosper plants and animals." *The Book of Rites* goes "unicorn, phoenix, tortoise and dragon are four paragons in the world." *Compendium of Materia Medica* says that dragon imitates nine animals on him, they are camel's head, deer's horns, rabbit's eyes, cow's ears, snake's neck, clam's stomach, carp's scales, hawk's paws, tiger's palms eighty-one scales totally on the back, nine times nine, the biggest number in China. There are whiskers on both sides of mouth, a bright pearl under the jaw, inversing scales under the throat, and mounts on top of the head".

The pattern of dragon looked big and fierce before Qin Dynasty, similar to reptile. Then during Qin and Han Dynasties, it turned to be a beast with four legs walking like a tiger or a horse. While in Sui and Tang Dynasties, the mouth was much wider, legs were longer, tail looked like a snake. In Song Dynasty, dragon pattern was similar to the previous one, but the jaw was cocking-up. In Yuan Dynasty, its hair appeared streaming in the wind; leg muscles and bones became visible. In Ming Dynasty, muscles and bone got protruding; the hair standed upright; whiskers curve out or in; and five-finger claws looked like windmills. At last, toward Qing Dynasty, hair growed wild over the head; saw-teeth cheek appeared; the tail became leave-shaped. The dragon pattern thus presented on the furniture, particularly seen in the imperial furniture decorations, and most demonstrated the majestic and grand of the imperial power.

综述 · 清乾隆时期木雕工艺

上图：缠枝纹
Above: Intertwined (Scrolled Grass) Pattern

下图：松竹梅
Below: Pine, Bamboo, and Plum Blossom

两个带梅字的植物在寒冬开花，但这两者都不是梅花，前者腊梅科，后者山茶科。

之所以将梅花扯进"岁寒三友"，大概是文人的喜好，梅花花先叶而开，老梅枝干遒劲而入画。因在早春开花，偶遇倒春寒，则是梅花香自苦寒来。

南宋末年有一个叫赵孟坚的画家，名气没有他的兄弟赵孟頫那么大，大概因为赵孟頫的"頫"字难认，很多人甚至还以为这个赵孟坚就是赵孟頫，只是写错了一个字。赵孟坚画过一幅松、竹、梅在一起的图画，名为"岁寒三友"，算是较早而直接的"岁寒三友"称呼。

有一个说法说岁寒三友之名最早来源于苏轼，他在被贬黄州时，一个大雪天，黄州知州来看他，觉得他的生活环境太冷清，但苏轼却回答他说"风泉两部乐，松竹三益友"。苏轼说了三友，并没有梅花。

将植物配一起，表达一下文人气质的，后来还有梅兰竹菊，在明朝给取了"四君子"之名，因为黄凤池辑《梅竹兰菊四谱》而来，如松竹梅一样，兰菊也很入画，就拉到了一起。

有了三友和四君子之后，解决了不少装饰设计的问题，若三面则松竹梅，若四面则梅兰竹菊。

这些植物还常被揉捏到一起，设计成一种装饰符号来表达一定的意义，若顽强、长寿往往是三友，清高、傲骨则四君子。甚至后来三友图还有了传宗接代的意思，仅仅因为竹之"笋"与"孙"同音。

乾隆花园建筑室内装修装饰大量采用松、竹、梅题材。

## 木雕用材

### 1.紫檀

属豆科植物，主要产于南洋群岛的热带地区，其次是柬埔寨、越南和泰国的东南部。紫檀木质坚硬，纹理纤细，变化无穷，色泽高雅，显

**Intertwined (scrolled grass) pattern**
Intertwined pattern is auspicious, also called "Longevity Vine", which implies good fortune. Vine grows long and continuously, which indicates endless, tender and wistful, fine and vivid. Intertwined pattern used for decoration originated from Han Dynasty, widely spreaded in the Northern and Southern Dynasties, Sui and Tang Dynasties, and Song, Yuan, Ming, Qing Dynasties afterwards.

Intertwined pattern often appears together with peonies, grapes, lotuses, and creatures which are called intertwined-peony, intertwined-grape, intertwined-lotus, and intertwined-creature, that is, intertwined pattern together with human, animals, birds. The intertwined pattern is seen on the furniture of Ming and Qing Dynasties decorations, but quite graceful and exquisite. The leaves became symmetrical, matching with components of the furniture.

**Pine, bamboo, and plum blossom**
Chinese are fond of pine, bamboo, and plum blossom because of the implications they own. They grow green and blossom in cold winter and symbolize brave, strong and righteous.

Chinese favor pine, bamboo and plum blossom as "three friends"; most literati associate themselves with the plants to demonstrate their bravery and perseverance. The plum blossom flower smells nice after experiencing the cold and harsh weather, so an old Chinese saying tells that everything goes well after going through all the difficulties in life.

A famous painter in Southern-Song Dynasty, named Zhao Mengjian, once painted pine, bamboo and plum blossom, titled *Three Friends in Cold Winter*, which is considered as the earliest naming for Three Friends in Cold Winter. Besides, famous poet in Northern-Song Dynasty Su Shi wrote a poem about Three Friends to express the hardships he had gone through while being demoted.

In Ming Dynasty, Huang Fengchi, compiled a book *Plum*

得静穆沉古。紫檀的特点一是密度大，入水即沉，不仅坚固耐用，还能抗腐蚀；二是生长缓慢，紫檀是常绿亚乔木，非经数百年不能成材。

**2.花梨木**

花梨木有两种，明代黄省曾在《夕阳朝贡典录》中介绍：其一为"花榈木"，乔木产于我国南方各地；其二为"海南檀"称为"黄花梨"。木质不静不喧，纹理或隐或现，生动多变者为花梨木的上品。

花梨木可分为新、老两类。老花梨木心材红褐至深红褐或紫红褐深浅不均，边材灰黄褐或浅黄褐色；新花梨木心材与边材区别不明显，纹理不如老花梨木生动，变化小，给人以柔和、文静的感受。

**3.鸡翅木**

鸡翅木又称杞梓木，比紫檀、花梨木更为稀少。因纹理似鸡的翅膀而得名。清初文学家屈大均在《广东新语》中介绍，鸡翅木有的白质黑章，有的色分黄紫，斜锯木纹呈细花云。籽为红豆，因此兼有"相思木"之名。

**4.铁力木**

铁力木产于我国广东、广西两地。木材质地坚硬，色泽纹理与鸡翅木相差无几，很难分辨。铁力木是常绿乔木，树干直立高可十余丈，直径达丈许。

**5.榉木**

榉木属榆科，落叶乔木。据陈嵘《中国树木分类学》中记载：榉木产于江、浙者为大叶榉木，别名"榉榆或大叶榆"，木材坚致，色纹并美，老龄者带赤色，名为"血榉"。

**6.红木**

红木在硬木中木质仅次于紫檀。老红木近

---

Blossom, Orchid, Bamboo and Chrysanthemums, *Four Plants*. Hence literati implied their bravery, perseverance, pureness and lofty with sweet plum, bamboo, orchid and chrysanthemums, named Four Friends.

Three Friends are found a lot in the interior decorations in Qianlong Garden architectures.

## Woodcarving material

### Red sandalwood
Red sandalwood is a member of the leguminosae family, and the best among hardwood. The wood mainly originates from the tropical forests, Cambodia, southeast of Vietnam and Thailand. The dense structure and grain of red sandalwood makes it extremely durable, so heavy that it sinks in water. Over 100 years, red sandalwood darkened and developed a fine surface patina. That's why red sandalwood was served for the imperial household in ancient China.

### Padauk wood
There are two types of padauk wood; one is from the south of China; the other is from Hainan Island, named huanghuali.

Padauk wood has old and new categories. The old wood is darker in color, maroon or brown red with quite clear grain, looking nice; while the new wood isn't, but looks soft and gentle.

### Chicken-wing wood
Chicken-wing wood, is dark brownish purple in color, with light and dark grains interleaved. The grain resembles a feather and is said unusual and beautiful. The seed is red, so called Acacia solid wood.

### Ceylon ironwood
Ceylon ironwood grows in Guangdong and Guangxi provinces. It is dense in structure; grain and color resemble chicken-wing wood. Ceylon ironwood is evergreen arbor, usually tens of meters high and meters in diameter.

紫檀但色暗，新红木颜色黄红，有花纹，年轮纹为直丝状，棕眼比紫檀木大。

**7.黄杨木**

黄杨木呈蛋黄色，非常悦目，木纹有直有斜，有光泽，木质光滑，结构细致，虽坚实但吃刀，易加工。

**8.楠木**

楠木属樟科，常绿乔木。《博物要览》记载：楠木有三种，一曰香楠，二曰金丝楠，三曰水楠。南方省多香楠，木微紫而清香，纹美。金丝楠出于川涧中，木纹有金丝，向明视之，镝烁可爱。楠木之至美者，向阳处结成人物山水之纹。水楠木色清而木质甚松，如水楠之类，唯可做木匾。楠木纹理直而美，耐水，切面光滑，有香气，属珍贵木材，尤其金丝楠现已罕见。

紫檀
Red Sandalwood

花梨木
Padauk wood

鸡翅木
Chicken-wing wood

铁力木
Ceylon ironwood

**Beech wood**

Beech wood, grows in Jiangxi and Zhejiang provinces, one of elm species, dense in structure, and grain looks nice.

**Mahogany**

A type of hardwood, Mahogany in China, is quite similar to red sandalwood in color and grain. It is difficult to distinguish them.

**Boxwood**

Boxwood is very nice in color, yolk yellow, grain is straight or oblique. It is so dense in structure and easy to be carved.

**Nanmu**

Nanmu is evergreen arbor, has three types according to *A Survey of Natural Science*. They look different since they grow in different geographic area. Nanmu's grain is straight, and it is endurable in water; tangent plane is quite smooth, and smells nice, regarded as precious wood.

榉木
Beech wood

红木
Mahogany

黄杨木
Boxwood

楠木
Nanmu

## 乾隆花园室内装修工艺特点

**1.竹木雕刻／镶嵌**

在雕刻木材之前，首先涉及木材种类的选取、对材料进行基本的处理。待"初坯"完成后，就正式进行雕凿了。当然整个过程还分解为画、耕、打点（铲、磨）、净地等工序。

木雕和砖雕一样，也有"拷活"。所谓"拷活"，在砖雕中就是将砖与砖重叠起来，一块砖拷在另一块上，以此对需要的部位进行更加立体的雕刻，使作品生动。

倦勤斋内檐装修的经典之一在于大面积运用竹簧雕刻工艺于仙楼上下槛窗下镶嵌的紫檀木壁板上。通过跟工匠、工艺美术大师的交谈，了解到倦勤斋下层东五间、上下两层仙楼的木壁板雕刻选用了不同的材料，涉及竹的各个部分，包括竹片、竹青、竹黄以及黄杨木等。运用竹黄雕刻的主要位置是山和远处（上层）的树；黄杨木雕刻则用在近处（下层）高大的树上，使得树干、树枝分外立体；而百鹿更是用竹片拼接后其上再贴竹黄，利用竹片、竹黄本身的纹理、管甬雕刻而成，甚至鹿毛都可以触及。回到正题，这里用到"拷活"这一特殊雕刻工艺的就是竹黄雕刻。处理之后的竹黄薄如纸，只能浅雕，而且纹理很直，需用刻刀刻出基本形状（不能用剪，一剪就自动裂开）。为了表现出一层次感，在雕刻前往往于其下垫层竹片（厚度在3毫米左右），竹片和竹黄间用鱼鳔胶黏结，这种结合比砖与砖连接处理起来更简单、稳固。待鱼鳔胶干后，就可以按程序进行画、耕等步骤了。

硬木镶嵌的群板、绦环板，先在板上镶嵌竹丝，再刻出凹槽（凹槽深于竹丝层，会刻到底板内），再将乌木口线嵌入。床挂面、罩楣挂楣的硬木（一般是黄杨木）雕刻的镶嵌也须在底板上刻出与镶嵌物形状一致的凹槽，再用鱼鳔胶黏合。

### Interior decoration characteristics of Qianlong Garden

**Bamboo and woodcarving and inlay**

The first step of carving is material selection and shaping; next is the primary carving; the work is completed after drawing, graving, scraping, grinding, cleansing and so on.

The highlight of Juanqinzhai (a Studio of Exhaustion form Diligent Service) is the interior decorations, where bamboo and woodcarving and inlay are found the refined ones among them. The One Hundred Deer and One Hundred Birds on the lower part of partitions in Juanqinzhai are typical artistic works of bamboo and woodcarving and inlay rather than masterpieces of traditional Chinese craftsmanship. The materials for the partitions are different parts of bamboo such as bamboo skin, green skin, inner skin, boxwood and so on. Inner bamboo skin carving is for the remote mountains and trees, boxwoodcarving for the closer big trees, so the trunks, branches and leaves look tridimensional. And the deer are carved with bamboo skin overlaying inner skin and wood woody xylem, so the hairs of deer can be seen due to the structure and grain of bamboo. As for inner bamboo skin, it is as thin as sheet of paper, hard to carve, so bamboo skin is adhered under it in order to appear picture layered.

Inlays are found on the middle and lower parts of the screens in Juanqinzhai. Bamboo filament marquetry is inlayed for the background in certain pattern, and then ebony wood strips are stuck into the grooves in the background. So do panels, partitions and other inlayed articles.

**Bamboo filament marquetry**

The eaves-hung fascia, partitions inside Juanqinzhai are inlayed with bamboo thread marquetry in 卍 pattern, elegant. Each unit is pieced together with thirteen bamboo filaments, appearing tridimensional under light. Each piece of filaments is made with

## 2.竹丝镶嵌

看似平整的挂檐板、绦环板、群板，细看都是细细竹丝镶嵌的绵延"卍"字纹，儒雅不凡。每个单元每条"卍"字边由13根深浅相间的竹丝拼合，光线充足的时候，板面分外立体。从残损、脱落的部位看，在拼合之前工匠是有布线这道工序的，底板上留下深浅不一的痕迹。还有就是纹饰按单元格拼贴的技巧。匠师自制的卡尺，同一卡尺上分几种规格（按长度不同划分）的卡槽，一端固定，在选择尺度后，在另一端用锯齿规范截取。这些竹丝的长度是在这样一种磨具的帮助下确定的，再经过布线排列，一组一组贴上去，严丝合缝。

## 3.画珐琅镶嵌

画珐琅器，称"洋瓷"。从清代蓝滨南在其《景德镇陶录》一书中对画珐琅器的描述可知，画珐琅器是以金属铜做器骨（胎），用五颜六色的瓷粉（珐琅釉）烧制而成。画珐琅器的制作方法是：先在已制成的红铜胎上涂施薄薄的下层白色珐琅釉，入窑烧结，并使其表面光洁平滑，然后以单色或多彩的珐琅釉料按照图案纹饰设计要求，绘制花纹图案，再经入窑焙烧显色而成，富有绘画趣味，因此，也有人称之为"珐琅画"。符望阁内的画珐琅镶嵌得到了大量运用。一般的隔扇槅心棂条镶嵌、横披隔扇屉心的棂条镶嵌用的是瓷胎的画珐琅；而罩槅组合的栏杆立柱扶手部位用的才是铜胎。花纹以西番莲、宝相花、蝙蝠为主，辅以卷草、夔纹，吉祥、华贵。

## 4.錾铜镶嵌

这种工艺保存的状况最好，基本没有开裂、脱落的迹象，应该与铜的物理性能（比如延展性）密切相关，且金属本身含水量很小，几乎不存在南北方气候差异导致的形变、脱落的情况。

self-made caliper, craftsmen pierce bamboo through the hole to cut bamboo filaments thin enough they required, then they are arranged into unites and adhered one by one to the panel or partition neatly on the basis of designed pattern.

**Painted enamel inlay**

Painted enamel is called foreign porcelain. Lan Binnan gave description of it in his book in Qing Dynasty; painted enamel is taken copper as the base, painted with colors of porcelain cement before firing in the kiln. Painted enamel is made in the following procedure: firstly, paint a thin layer of white porcelain cement on the red copper base; then fire the article to smooth the surface and paint pictures designed with color porcelain cement; thirdly, fire again for color development. The manufacture of painted enamel is not only firing porcelain but also painting, so regarded as enamel painting. Painted enamel inlay is found a great deal in Fuwangge of Qianlong garden. It is set on parts of partitions, some are copper-based, and some are porcelain-based enamel. The painting is mostly passionflower, bats and scrolled and kui-dragon pattern to present good fortunes and luxury.

**Chiseled copper inlay**

Chiseled copper is so far remained the best among all the decorations, few cracks, falling, much to do with the physical properties of copper, for example, ductility and low moisture, unable to deform or break under extreme weathers between the south and the north.

**Mother-of-pearl inlay**

Mother-of-pearl inlay works with shells of clam, abalone, spiral, etc. After cutting, grinding and piecing together and set them into lacquer base to form patterns, then repaint and polish. Two types of mother-of-pearl inlays are observed in Fuwangge: one is thin-dotted; the other is thick-sliced. Thin-dotted is manufactured by lacquer craftsmen in Yangzhou, Jiang Qianli, at the end

### 5.螺钿镶嵌

精选河蚌、鲍鱼贝、夜光螺等优质贝壳作为原料，经过磨制后，做成图案，拼贴、镶嵌于漆坯上，在经过髹漆、抛光等步骤制作而成。从符望阁的内檐装饰中可以看到软、硬两种螺钿的镶嵌工艺。软螺钿镶嵌主要是点螺，为明末扬州漆器匠师江千里所创。因其磨制的螺钿片薄如蝉翼，切割成的点、丝、片均细如秋毫，需用特制工具点嵌于漆坯上，故名。点螺漆器纹样精细灵巧，色彩天然，随光变幻，远看似浮雕，有强烈立体感，近看却平滑如镜，光可鉴人，是漆器中名贵的品种。如此精美的工艺用在挂檐底板装饰上衬托雕漆、博古，可谓匠心独运、气魄非凡。硬螺钿磨制得厚一些，又分凸嵌和平磨两种，一般用于片状构图的镶嵌，如花瓣，梅花、茶花都有；亦有条状的，如西面北次间内朝西、朝南两座床罩床挂面的夔龙缠枝花卉的夔龙纹就部分运用硬螺钿凸嵌。

### 6.玉石镶嵌

玉片嵌在木料上，材料的性质差异使得只用鱼鳔胶稳固性是得不到保障的，于是匠师们想出了用铜丝固定的方法。把雕刻好的玉片背后（一般中心位置）钻上对应的小孔，再用细铜丝从木料背后穿过玉片相通的孔再回穿过木料，最后于棂条背后系紧。用铜丝而不用铁丝或者其他金属丝，一来是因为铜的稳定性高，不似铁易生锈腐蚀折断，二来用铜从造价上也能够接受。除了槅心棂条、绦环板、群板等的规则几何纹样（或圆环、或菱环）镶嵌外，玉片也和其他珍贵宝石一起用在挂楣、床炕挂面的花卉、果实、器物的镶嵌中。

### 7.宝石镶嵌

其他以玛瑙、绿松石、青金石、芙蓉石、孔雀石等珍稀宝石镶嵌的植物花卉、果实、器物、纹样等。

of Ming Dynasty. The pear shell is ground as thin as cicada's wing, cut into tiny pieces as small as dots and hairs, so-called thin-dotted inlay. The pattern of mother-of-pearl inlay is so fine that the color changes under light resembling painting, appearing tridimensional but smooth and reflecting as a mirror, master piece of lacquer articles. Such refined art crafts are applied in the eave-hung fascia to create the magnificence of lacquer carving and antiques inlay. Thick-sliced mother-of-pearl inlay is ground thicker than the thin-dotted one, ground into flat and convex surface for sliced objects and flower petals. For example, kui-dragon pattern is smoothed mother-of-pearl inlay, and the flower petals like sweet plum and camellia are inlayed in the bed-sided boards in Fuwangge.

**Jade inlay**

Sliced jade is adopted in the inlay decorations as well; the difference among materials makes fish glue lose its firmness; craftsmen came up with copper wire to fasten. That is to drill a tiny hole at the back of the carved jade slice and a hole into the wood, copper wire runs through the holes, and fastens tightly to the wooden lattice behind. Carved jade and precious stones are inlayed on the screens, partitions, bed-sided boards, articles, etc. to express flowers, fruits, objects, and so on.

**Precious stone inlay**

The precious stone inlays are agate, turquoise, malachite, lazurite, quartz, etc.

**Lacquer carving technique**

Lacquer carving technique originated from Tang Dynasty. Lacquer is carved into a designed pattern or picture on an article as a result lacquer is painted one after another for tens even hundreds of layers. Lacquer carving can be distinguished into red, beige, color, etc, and red lacquer takes the most. The lacquer panels found in Fuwangge are carved red lacquer of the character "寿" indicating longevity.

Inlays are applied to almost all interior decorations

**8.雕漆工艺**

雕漆也叫剔红,技艺始于唐代,是在髹涂数十层至数百层的漆面上,雕刻以各种纹样为装饰的漆器。一般以锦纹为地,花纹隐起,喜庆华美,典雅庄重。雕漆按其漆色分,有剔红、剔黄、剔彩等,以剔红为主。阁内除东面明间(朝东),东、西次间进深方向的罩槅挂楣的开光用剔红(回纹)外,隐藏在东面明间朝西一侧的挂楣上,剔红的开光内剔红的"寿"字凸显工匠的精湛技艺以及吉祥喜庆的寓意。

在乾隆花园各建筑内檐装修中,"镶嵌"的手法贯穿始终。无论是竹丝、玉片、螺钿、画珐琅、錾铜、珍稀宝石、瓷片、竹木雕刻都镶嵌于挂楣、横披和槅心棂子、绦环板、群板、床挂面等的底坯(或漆底、或硬木、或竹丝镶嵌)之上用做装修装饰。

竹木雕刻 / 镶嵌
Bamboo and Wood Carving and Inlay

竹丝镶嵌
Bamboo Filament Marquetry

画珐琅镶嵌
Painted Enamel Inlay

錾铜镶嵌
Chiseled Copper Inlay

of partitions, screens, and panels in Qianlong Garden. Inlays include materials of bamboo filament, jade, mother-of-pearl, enamel, copper, stone, porcelain, bamboo and wood carving, lacquer and so on.

螺钿镶嵌
Mother-of-pearl Inlay

玉石镶嵌
Jade Inlay

宝石镶嵌
Precious Stone Inlay

雕漆工艺
Lacquer Carving Technique

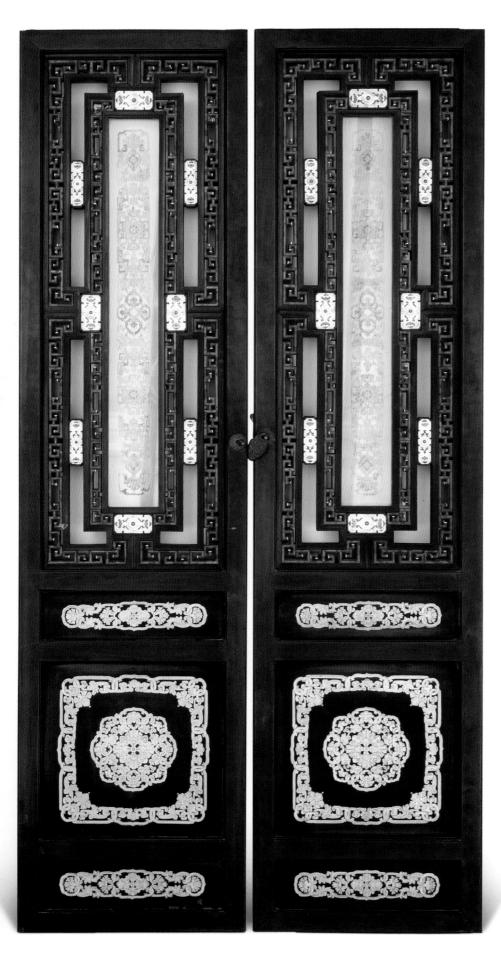

## 乾隆花园内清代宫廷家具的风格及特点

在清代家具中，康熙、雍正、乾隆三个时期制作的家具代表着典型的清式风格。更确切地说，清代家具新的做法、造型、装饰在雍正时已经具备，至乾隆年达到高峰，以后逐渐下降。

清式家具与明代家具的共同特点是都以优质木材制成。而在造型艺术及风格上却有很大差异。清式家具的特点主要表现在以下几个方面：①用材厚重，家具的总体尺寸较明代宽大，相应的局部尺寸也随之加大；②装饰华丽，表现手法主要是镶嵌、雕刻及彩绘等技法。清式家具给人稳重、精致、豪华、艳丽的感觉，和明式家具的朴素、大方、优美、舒适形成鲜明的对比。

清式家具和明式家具相比，整体不像明式家具那样以朴素大方、优美舒适为标准，而是以厚重繁华、富丽堂皇为标准，因而显得厚重有余，俊

## The style and characteristics of the furniture of Qing Dynasty in Qianlong Garden

In Qing Dynasty the furniture made during Kangxi, Yongzheng and Qianlong periods represent typical Qing style. To be precise, new technique, style, and decorations had formed their shape at Yongzheng time, got to their climax at Qianlong time, and then declined gradually.

Precious and fine wood as furniture material is the similarity between Ming style furniture and Qing style furniture. However, the style and shape are distinguished. Qing style furniture is firstly heavy and bigger in size; secondly, decorations play an important role, such as inlay, carving, and painting. Therefore, it looks stable, exquisite, splendid and gorgeous; while Ming style furniture is plain, simple, elegant and comfortable.

秀不足，给人沉闷笨重之感。但从另一方面说，由于清式家具以富丽、豪华、稳重、威严为准则，为达到设计目的，利用各种手段，采用多种材料、多种形式，巧妙地装饰在家具上，效果也非常好。所以，清式家具和明代家具各有千秋，都是中国家具艺术中的精品。

清式家具的产地主要有广州、苏州、北京三处，它们各自有着迥然不同的地方风格，被称为代表清代家具的三大名作。其中，又以广作家具最为突出。

**1.广作家具**

明末清初，海外贸易相对繁荣，广州因其特定的地理位置，成为中国对外贸易和文化交流的重要门户。加之广东又是贵重木材的主要产地，南洋各国的优质木材也多由广州进口，因此，制作家具的材料比较充足。得天独厚的有利条件，赋予了广作家具的艺术风格。

[明] 黄花梨四出头官帽椅
Ming Dynasty Padauk Wood Armchair with Four Protruding Ends

To compare with Ming style furniture, Qing furniture takes being luxuriant and exquisite as standard, so it is heavy rather than delicate. On the other hand, the decorative effect is good enough since so many techniques, materials, designs are applied to use. Ming style furniture and Qing style furniture both are choices in Chinese furniture.
Qing style furniture is mainly manufactured in Guangzhou, Suzhou and Beijing. They differ from one another, representing local style, and they are three representative furniture manufacturing workshops. Guangzhou style seems the best among them.

**Guangzhou-style furniture**
By the end of Ming Dynasty and the beginning of Qing Dynasty, overseas trade got flourishing in Guangzhou due to its special location. Guangzhou became an important port for foreign trade and cultural exchange. In addition, Guangdong was then the main origin of precious wood which was imported by South Asian countries. Therefore, adequate resources of materials

[清] 紫檀透雕卷草纹藤心圈椅
Qing Dynasty Red Sandalwood Chair with Openwork Carving in Scrolled Grass Design

广作家具的特点之一是用料粗大充裕。一件家具的各部构件，尤其是腿足、立柱是主要构件，不论弯曲度有多大，一般不用拼接做法，而习惯用一块整料挖成。其他部位也大体如此，所以广作家具用料大都比较粗壮。为讲求木性一致，广作家具大多使用一种木料制成。通常所见的广作家具，或紫檀，或红木，皆为清一色的同一木质，决不掺杂别种木材。而且广作家具不加漆饰，使木质完全裸露，让人一看便有实实在在、一目了然之感。

广作家具的特点之二是装饰花纹雕刻深湛、刀法圆熟、磨工精细。其雕刻风格在一定程度上受到西方建筑雕刻的影响。雕刻花纹隆起较高，个别部位近乎圆雕。加上磨工精细，使花纹表面光滑如玉，丝毫不露刀凿痕迹。虽然雕刻较深，用手摸时，却有圆滑柔和之感，而衬地表面亦平滑如镜。在板面图案纹理复杂、铲刀受阻的情况下，要把地子处理得这样平，在当时手工操作的条件下，是很不容易的。这种雕刻风格，在广作家具中较为突出。

广作家具的装饰题材也受到了西方文化艺术的影响。明末清初之际，西方的建筑、雕刻、绘画等技术逐渐为中国所应用。自清代雍正至乾隆、嘉庆时期，模仿西式建筑的风气大盛。除广州外，其他地区也有这种情况。如在北京兴建的圆明园，其中就有不少建筑从形式格局到室内装修，都是西洋风格。为装饰这些殿堂，清宫每年除从广州定做、采办大批家具外，还从广州挑选优秀工匠到皇宫，为皇家制作与这些建筑风格相协调的中西结合式家具。即以中国传统做法制成器物后，通常是一种形似牡丹的花纹，也有称为"西番莲"的。这种花纹线条流畅，变化多样，可根据不同器形而随意伸展枝叶，其特点是多以一朵花或几朵花为中心向四周伸展枝叶，且大多上下左右对

offer a priority to develop Guangzhou-style furniture. An obvious character of Guangzhou-style furniture is the big size. Every component like legs, poles are made of an entire piece of wood, they wouldn't be pieced together no matter how much they bend. Other parts of the furniture would resemble this. Guangzhou-style furniture emphasizes the integrity of material; the whole piece of furniture will only apply one type of wood. Red sandalwood or mahogany furniture is made of pure red sandalwood or mahogany. No one will add any other wood to substitute for it. Guangzhou-style furniture doesn't apply lacquer, but bare wood usually.

The second character of Guangzhou-style furniture is that the carving is deep, round and meticulous. The carving style was influenced by the western architectural engrave to some extent. Some parts are similar to rounded engrave, so it feels smooth and evasive. The base is chiseled so smooth and flat as mirror, quite challenging at the time, especially when the carving pattern is complicated, chisels can't reach directly. Nevertheless, this kind of carving is rather popular in Guangzhou-style furniture.

Carving pattern of Guangzhou-style furniture was influenced by the western culture and art. Western architecture, engrave, painting were introduced to China by the end of Ming Dynasty and at the beginning of Qing Dynasty. Imitating western architectural style came to be that popular since Yongzheng period to Qianlong and Jiaqing periods. Other places followed the trend including Guangzhou. The construction of Yuanmingyuan in Beijing is an example. A great number of architectures and interior decorations raise the view of western style. A great deal of the furniture was ordered from Guangzhou every year in order to decorate the imperial palaces, a lot of craftsmen were also selected to the Imperial Court to manufacture furniture, which combined Chinese styles with Western styles to match with the architectural style, as well. So-called passionflower is similar in appearance with peony, the winding vine and leaves circle around the flower extending and spreading out to the shape of the article, up and down, right and left symmetrically, and the beginning meets the end

称。如果装饰在圆形器物上，各面花纹衔接巧妙，很难分辨它们的首尾。

广作家具除装饰西洋花纹外，也有相当数量的传统纹饰，如各种形式的海水云龙纹、海水江崖纹、云纹、凤纹、夔纹、蝠纹、缠枝或折枝花卉纹以及各种花边装饰等。有的广作家具上中西两种纹饰兼而有之，也有的广作家具乍看都是中国传统花纹，但细看起来，或多或少地总带有西式纹饰的痕迹。这为我们鉴定是否为广作家具提供了重要的依据。

广作家具的镶嵌艺术也很发达。这种艺术风格主要表现在雕刻和镶嵌的艺术手法上。镶嵌作品多为屏风类和箱柜类家具，原料以象牙、珊瑚、翡翠、景泰蓝、玻璃画等为主。

提到镶嵌，人们多将其与漆器联想在一起。原因是我国镶嵌艺术多以漆做地，而广作家具的镶嵌却很少见漆，这是有别于其他地区的一个明显特征，镶嵌内容多以山水风景、树石花卉、走兽飞禽、神话故事以及反映现实生活的风土人情为题材。

广州还有一种以玻璃画为装饰材料的家具，以屏风类家具为常见。玻璃油画就是在玻璃上画的油彩画，于明末清初由欧洲传入中国，首先在广州兴起，曾形成专业生产。中国现存玻璃油画，除直接从外国进口外，大多由广州生产。它与一般绘画画法不同，是用油彩直接在玻璃的背面作画，而画面却在正面。其画法是先画近景，后画远景，用远景压近景。尤其是画人物的五官，要画得气韵生动，就更不容易了。在乾隆至嘉庆年间，曾形成专业生产。

**2.苏作家具**

苏作家具是指以苏州为中心的长江下游一带所生产的家具。苏作家具形成较早，举世闻名的明式家具即以苏作家具为主，它以造型优

unidentified when this pattern is applied on a round object.
There is still large amount of furniture decorated with traditional pattern. For example, it is found in types of sea and dragons, sea and rocks, clouds, phoenix, kui-dragons, bats, intertwined and scrolled branches, etc. Some have both western and Chinese styles, some appear to be Chinese, but more or less are western in some details, which provide evidence to recognize the style of furniture.

Inlays are noticeable in Guangzhou-style furniture. They are combined with carving to produce an artistic view. Inlays are applied mostly on the screens, boxes and cabinets with ivory, coral, jade, cloisonne, glass painting, etc.

As for inlays, they are usually connected with lacquer because material is often inlayed on the lacquer base; while Guangzhou-style inlays are not, which is a regional specialty for Guagnzhou-style inlay. The inlay contains landscapes, trees and flowers, birds and animals, legends and tales, even local life and custom.
Among the inlay materials, glass painting was found in the screen. Glass painting was first introduced into Guangzhou at the end of Ming Dynasty from Europe, developed to certain scale afterwards. The extant glass painting in China is mainly produced in Guangdong province apart from the imported. Glass painting is different from the painting on paper or silk. It is painted on the back side of the glass. The view close to people is painted before the view away in distance. Figure painting, faces are particularly hard to paint. However, glass painting was then in Guangzhou extraordinary professional.

**Suzhou-style furniture**
Suzhou-style furniture refers to the furniture manufactured in the places centered on Suzhou along middle and lower Yangtze River. Suzhou-style furniture came into shape early. The well-known Ming style furniture is mainly Suzhou-style. It was favored by people for its elegant shape,

美、线条流畅、用料结构合理、比例尺寸适度等特点和朴素大方的风格博得了世人的赞赏。进入清代以后，随着社会风气的变化，清代苏作家具也开始向烦琐和华而不实转变。这里所讲的苏作家具，主要是对清代而言的。

苏作家具用料节俭是其特点之一。通常所见苏作椅子，除主要承重构件外，多用碎料攒成。椅腿直面以外的所有装饰全部用碎料粘贴，包括回纹马蹄部分所需的一小块薄板。椅面下的牙条也较窄较薄，座面边框也不宽，中间不用板心而用席，这样节省了许多木料。面上的靠背扶手多采用拐子纹装饰。拐角处做榫拼接。这种纹饰用不着大料，甚至连拇指大小的木块都可以派到用场，足见用料之节俭。

苏作家具的大件器物还时常采用包镶手法，即用杂木做成骨架，外面粘贴硬木薄板。这种包镶做法，费力费时，技术要求也较高。好的包镶家具不经过仔细观察或用手掂一掂，很难看出是包镶做法。为了不让人看出破绽，通常把拼缝处理在棱角处，使家具表面木质纹理保持整洁，既节省了材料，又不破坏家具本身的整体效果。为了节省材料，在制作家具时，还常在暗处掺杂其他柴杂木。这种情况，多表现在家具里面的穿带上。现在故宫博物院收藏的苏作家具，很多都有这种情况。苏作家具都在里侧油漆，目的在于避免受潮，保证木料不变形，同时也有遮丑的作用。

苏作家具的镶嵌和雕刻艺术主要表现在箱、柜和屏风类器物上。以常见箱柜为例，多以硬木做成柜框，框架之间起槽镶一块松木或杉木板，然后按漆工工序做成素漆面，漆面阴干后，开始装饰图案，先在漆面上描出画稿，再按图案形式用刀挖槽，将事先按图做好的各种质地的嵌软件镶进槽内，用胶粘牢，即为成器。苏作家具中的各种镶嵌也多用小块材料堆嵌，整板上面积雕刻成器的不多。常见的镶嵌

fluent line, comfortable structure, moderate size, simple and easy style. As the trend went in Qing Dynasty, it tended to be luxury and flashy.
One character of Suzhou-style furniture is material thrifty. For instance, a Suzhou-style chair, most components were pieced together except load-bearing parts. Decorating parts for legs were also patched together, so were horse-hoof foot in shape. Seat was not wood but rattan mat except the frame, and strips beneath seat were thin and narrow so as to save wood. Chair back and armrest were adopted rectangular spiral pattern with mortise and tenon joints at the corners because this design didn't need big pieces of wood. Therefore every piece of material makes its use, though one as small as a finger.

As for big piece of furniture or article, Suzhou-style furniture apply weed tree wood inside and hard or precious wood board wrapped up outside. This is what people consider baoxiang (common wood is wrapped with thin pieces of precious wood outside so as to appear luxuriously), this technique requires high skills. A well-wrapped piece of furniture can be hardly verdicted before observed with great care and weighted by hands. The patching parts usually happen at the edges and corners to avoid flaws. The outside surface appears smooth and integrates with one type of wood grain, whereas inside is painted with lacquer in order to prevent from damp and hide up the defect. This manufacturing technique is found in many pieces furniture collected by the Palace Museum.

Inlay and carving art are adopted in the producing of cases, cabinets and screens of Sunzhou-style furniture. Take cases for example, the structural frames are usually made of hardwood, sides are pine or cedar wood, and then cover with lacquer, draw pictures when lacquer getting dry, dig on the basis of picture to set mother-of-pearl, ivory, jade, and colorful stones into the ditches. Carving also takes a large amount of Suzhou-style furniture. Suzhou artisans thus made full use of every piece of materials; nothing would be wasted no matter how small it is.

材料多为玉石、象牙、螺钿和各种颜色的彩石，也有相当数量的木雕。苏作家具镶嵌手法的主要优点是可以充分利用材料，哪怕只有黄豆大小的玉石碎渣或螺钿碎末，都不会废弃。

苏作家具的装饰题材多取自历代名人画稿，以松、竹、梅、山石、花鸟、山水风景以及各种神话传说为主。其次是传统纹饰，如海水云龙、海水江崖、双龙戏珠、龙凤呈祥等。折枝花卉亦很普遍，大多借其谐音寓意一句吉祥语。局部装饰花纹多以缠枝莲和缠枝牡丹为主，西洋花纹极为少见。一般情况下，苏作的缠枝莲纹和广作的西番莲纹，是区别苏作和广作的一个特征。

**3.京作家具**

京作家具一般以清宫造办处所制家具为主。造办处内有单独的广木作，由广东优秀工匠担任，所制器物较多体现着广作风格。因此造办处在制作某一件器物前都必须先画样呈览，经皇帝批准后方可制作。在这些御批中，经常记载着这样的事：皇帝看了画样后，觉得某部分用料过大，及时批示将某部分收小一些。久而久之，形成了京作家具较广作家具用料较小的特点。在造办处普通木作中，多从江南广大地区选招工匠，做工趋向苏作。不同的是他们在清宫造办处制作的家具较苏作家具用料稍大，而且没有掺假的现象。

从纹饰上看，京作家具又较其他地区独具风格。它从皇宫中收藏的古代铜器、玉器及石刻艺术上吸取素材，巧妙地装饰在家具上。明代就已开始在家具上雕刻古铜器、古玉器纹饰，清代在明代的基础上使用更加广泛了。明代常用装饰头案的牙板和案足间的挡板，而清代则在桌案、椅凳、箱柜上普遍应用。明代多雕刻螭纹（北京匠师称其为螭虎龙或草龙），而清代则是夔纹、夔凤纹、拐子纹、螭纹、虬纹、饕餮纹、兽面纹、雷纹、蝉纹、勾卷云纹等无所不有，根据家具的不同造型特点，而施以各种不同形态的纹饰，显示出各式古色古香、文静典雅的艺术形象。

Suzhou-style decoration subjects contain elite's paintings, most are pine, bamboo, sweet plum, rocks, flowers, birds, landscapes, and legends. Traditional patterns such as sea and dragon, dragon and phoenix take the second place. Branches and flowers are common. They symbolize good fortunes. Intertwined-lotus and intertwined-peony are essential for detailed decoration; however, western pattern is scarcely noticed. So, it is a factor to differ Suzhou-style from Guangzhou-style furniture, the former is intertwined lotus, and the latter is passion flower.

**Beijing-style furniture**
Beijing-style furniture is mainly made by the wood workshop of Imperial Household Department of Qing Dynasty. The best carpenters were from Guangdong, they manufactured Guangzhou-style furniture, but they had to submit the painting samples to the emperors for approval. The emperor mended to downsize some components to save material till he felt pleased. So in this case, Beijing-style furniture formed its shape, which was smaller than Guangzhou-style furniture. Other inferior wood workshop owned most carpenters from Jiangnan area, furniture tended to be Suzhou-style, but it was relatively bigger than those made in the South, though having no inferior wood hidden inside or wrapped up with hardwood outside.

Imperial furniture is unique to other styles in decoration patterns. Originated from Ming Dynasty, popularized in Qing Dynasty the decorative styles were influenced by ancient bronze, jade and stone carving articles collected in the imperial court. These patterns were applied on the side panels of the side tables in Ming Dynasty and tables, chairs, cabinets, etc in Qing Dynasty. They are caved chi-dragon, also called chihu-dragon or intertwined-dragon in Ming Dynasty, while kui-dragon, kui-phoenix, chi-dragon, qiu-dragon, glutton motive, monster-faced design, cicada design, scrolled cloud design etc, almost all types of patterns are found in the furniture of Qing Dynasty. Furniture was decorated according to its shape and style indicating differently, appearing the antique flavor.

Chapter II Drawings

第二部分·图版

Bamboo and Wood Carving

竹木雕刻

## 硬木宫扇底座
### Hardwood Court Fan Seat

❶ 底座顶部纹饰
Detail of the Decoration on Top Section of the Seat

❷ 底座细部卷草纹饰
Detail of Carved Scrolled Grass Decoration

❸ 底座须弥座束腰卷草纹饰
Detail of Carved Scrolled Grass Decoration

## 紫檀透雕龙凤盘长纹六角墩
### Red Sandalwood Hexagonal Stool with Openwork Carving in Dragon and Phoenix Design

❷ 墩壁边框肩部
Detail of the Joint between Top Rail and Leg

❸ 墩壁透雕龙头图案
Detail of the Side Panel of the Stool with Openwork Carving in Dragon Head Design

❹ 墩壁透雕凤头图案
Detail of the Side Panel of the Stool with Openwork Carving in Phoenix Head Design

**❶ 透雕龙凤盘长纹墩壁**
Detail of the Side Panel with Openwork Carving in Dragon and Phoenix Design

## 紫檀透雕云蝠纹六角墩
### Red Sandalwood Hexagonal Stool with Openwork Carving in Cloud and Bat Design

❷ 墩壁蝠纹细部 1
Detail of the Side Panel with Openwork Carving in Bat Design 1

❸ 墩壁蝠纹细部 2
Detail of the Side Panel with Openwork Carving in Bat Design 2

❹ 墩壁蝠纹细部 3
Detail of the Side Panel with Openwork Carving in Bat Design 3

❶ 透雕云蝠纹墩壁

Detail of the Side Panel with Openwork Carving in Cloud and Bat Design

## 紫檀透雕二龙戏珠四方委角墩
### Red Sandalwood Square Stool with Openwork Carving in Two-Dragon-Playing-with-the-Pearl

❷ 墩壁龙头纹饰
Detail of the Side Panel with Carving in Dragon Head Decoration

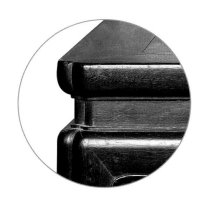

❸ 束腰
Detail of the Frieze

❹ 龟脚
Foot

❶ 透雕二龙戏珠墩壁
Detail of Side Panel with Openwork Carving in Two-Dragon-Playing-with-the-Pearl Design

## 紫檀雕花纹绣墩
### Red Sandalwood Garden Stool with Carved Decorations

❷ 墩壁西洋花雕刻细部 1
Detail of Carved Flower on
Stool Panel 1

❸ 墩壁西洋花雕刻细部 2
Detail of Carved Flower on
Stool Panel 2

❹ 墩壁西洋花雕刻细部 3
Detail of Carved Flower on
Stool Panel 3

❶ 紫檀雕花纹绣墩墩壁局部
Detail of Side Panel of Red Sandalwood Garden Stool

## 紫檀木雕花纹桃式机
### Red Sandalwood Stool with Carved Decorations

❶ 牙子虬纹雕刻细部 1
Detail of Qiu-dragon Patterns
Carving on the Apron 1

❷ 牙子虬纹雕刻细部 2
Detail of Qiu-dragon Patterns
Carving on the Apron 2

❸ 云纹足机腿
Detail of Stool Foot in
Cloud Motif

## 紫檀木绣墩
### Red Sandalwood Garden Stool

**❶ 开光内蝠纹雕刻细部**
Detail of Carved Bat Design on the Medallion

**❷ 开光内鱼纹雕刻细部**
Detail of Carved Fish Design on the Medallion

**❸ 墩壁蝠纹雕刻细部**
Detail of Carved Bat Design on the Stool Panel

## 鸡翅木六开光绣墩
### Chicken-wing Wood Stool with Seat Cushion

❶ 玄纹鼓钉
Detail of Carved Drum-nail Pattern Decoration

❷ 腹部椭圆形开光
Detail of Oval Side Panel

❸ 细部
Detail

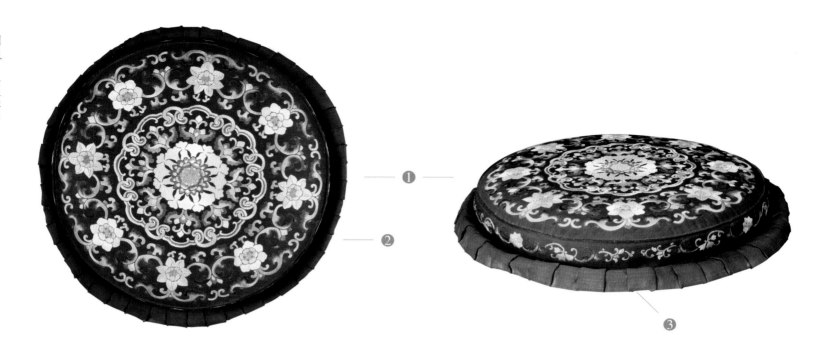

## 鸡翅木六开光绣墩——蓝缎垫
## Chicken-wing Wood Stool with Seat Cushion-Blue Satin Cushion

❶ 中心花卉图案
Detail of the Central Flower Pattern

❷ 如意云纹图
Detail of S-shaped Cloud Pattern

❸ 缠枝花卉图案细部
Detail of Intertwined Flower Pattern

**❶ 椅背雕刻细部**
Detail of the Carving on the Backrest

## 紫檀透雕卷草纹藤心圈椅
## Red Sandalwood Chair with Openwork Carving in Scrolled Grass Design

**❷ 卷草纹扶手**
Detail of Carved Scrolled Grass Armrest

**❸ 内翻马蹄**
Varus Horseshoe

**❹ 龟脚**
Foot

**❶ ❷ 松鹿山石雕刻靠背板楠木板心**
Nanmu Backboard with Carved Pine, Deer, Mountain and Stone Design

## 紫檀木嵌楠木雕花七屏式宝座及足踏
## Red Sandalwood and Nanmu-carved Seven-screen-pattern Throne with Foot Stool

**❸ 云龙纹帽子**
Detail of the Top with Carving in Dragon and Cloud Decoration

**❹ 牙条拐子纹雕刻**
Detail of the Rectangular Spiral Pattern Carving

**❺ 卷云足**
Foot with Carved Cirrus Cloud Pattern

楠木板心雕刻细部 1
Detail of Carved Nanmu Panel 1

楠木板心雕刻细部 2
Detail of Carved Nanmu Panel 2

楠木板心雕刻细部 3
Detail of Carved Nanmu Panel 3

楠木板心雕刻细部 4
Detail of Carved Nanmu Panel 4

楠木板心雕刻细部 5
Detail of Carved Nanmu Panel 5

楠木板心雕刻细部 6
Detail of Carved Nanmu Panel 6

❶ 靠背蝠磬图案雕刻
Detail of Carved Bats and Chime Stone Pattern on the Backrest

❷ 靠背夔龙图案雕刻
Detail of Carved Kui-dragon Pattern on the Backrest

## 紫檀雕夔凤七屏式宝座
### Red Sandalwood Seven-screen-style Throne with Carved Kui-phoenix Design

❸ 靠背雕刻细部
Detail of Carved Backrest

❹ 牙条如意纹雕刻
Detail of S-shaped Carving

❺ 卷云足
Foot with Carved Cirrus Cloud Pattern

宝座正面、侧面、背面
Front, Side and Back of the Throne

❶ 紫檀木嵌黄杨宝座局部
Detail of Red Sandalwood Throne with Boxwood Inlays

## 紫檀木嵌黄杨宝座
## Red Sandalwood Throne with Boxwood Inlays

❷ 搭脑西洋巴洛克式装饰
Detail of Carved Decoration in Baroque Style

❸ 牙板雕刻细部
Detail of Carved Apron

❹ 拱肩雕刻
Detail of Carved Spandrel

❶ 紫檀木剔黄宝座局部
Detail of Red Sandalwood Throne with Carved Yellow Lacquer Decorations

## 紫檀木剔黄宝座
## Red Sandalwood Throne with Carved Yellow Lacquer Decorations

❶ 床围雕刻细部
Detail of Carved Bed Bumper

❷ 束腰开炮筒式鱼门洞
Detail of Decoration on the Waist

❸ 内翻马蹄宝座腿
Detail of Varus Horseshoe Foot

图版 · 竹木雕刻

第二部分

075

❹ 扶手嵌玉木雕松竹梅图案 1
Armrest with Carved Pine, Bamboo and Plum Blossoms
Patterns with Jade Inlay 1

❺ 扶手嵌玉木雕松竹梅图案 2
Armrest with Carved Pine, Bamboo and Plum Blossoms
Patterns with Jade Inlay 2

❶ 椅背雕刻细部
Detail of the Carving on the Backrest

## 天然木罗汉床及足踏
## Natural Wood Arhat Bed and Foot Stool

❷ 床腿细部 1　　　　　　　　　❸ 侧部雕刻　　　　　　　　　❹ 床腿细部 2
Detail of the Leg 1　　　　　Detail of the Side Carving　　　Detail of the Leg 2

天然木罗汉床及足踏
Natural Wood Arhat Bed and Foot Stool

图版・竹木雕刻

❶ 椅背雕刻细部
Detail of the Carving on the Backrest

# 天然木椅
# Natural Wood Chair

❷ 椅腿细部 1  ❸ 椅腿细部 2  ❹ 扶手细部
Detail of the Leg 1   Detail of the Leg 2   Detail of the Armrest

❶ 桌腿细部 1
Detail of the Leg 1

## 天然木小圆桌
## Natural Wood Round Table

❷ 桌腿细部 2　　　　　　　❸ 桌腿细部 3　　　　　　　❹ 桌面边缘细部
Detail of the Leg 2　　　　　Detail of the Leg 3　　　　　Detail of the Top Edge

·图版·竹木雕刻·

## 紫檀木雕花纹半圆桌
### Red Sandalwood Semi-circular Table with Carved Decorations

❶ 牙条细部
Detail of Carving on the Apron

❷ 地盘透雕细部
Detail of Openwork Carving

❸ 腿足细部纹饰
Detail of Carved Decoration on Foot and Leg

## 紫檀木雕花纹六角香几
### Red Sandalwood Hexagonal Incense Stand with Carved Decorations

❶ 束腰西洋花细部雕刻
Detail of Carved Flower on the Waist

❷ 牙子西洋花雕刻
Detail of Carved Flower on Apron

**❶ 桌腿细部 1**
Detail of Carved Leg

# 紫檀木圆桌
# Red Sandalwood Round Table

**❷ 桌腿细部 2**
Detail of Carved Leg

❶ 透雕西洋花牙板细部
Detail of the Apron in Openwork Carving in Western Flower Design

## 紫檀透雕西洋花牙板长桌
## Red Sandalwood Table with Apron in Openwork Carving in Western Flower Design

❷ 透雕西洋花纹饰细部 1
Detail of the Openwork Carving in Western Flower Pattern 1

❸ 透雕西洋花纹饰细部 2
Detail of the Carving in Western Flower on the Apron 2

❹ 桌腿纹饰细部
Detail of the Pattern on the Leg

① 紫檀木桦木心长桌局部
Detail of Red Sandalwood Table with Birch Central Panel

## 紫檀木桦木心长桌
## Red Sandalwood Table with Birch Central Panel

② 桌腿与罗锅枨细部 1
Detail of Leg and Stretcher 1

③ 桌腿与罗锅枨细部 2
Detail of Leg and Stretcher 2

④ 内翻云纹马蹄
Detail of Varus Horseshoe in Cloud Pattens

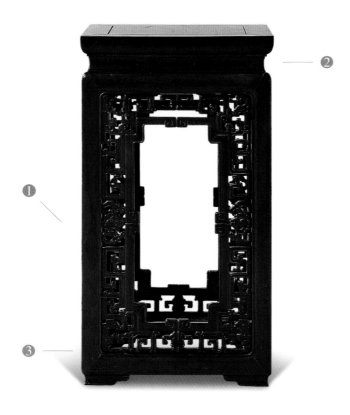

## 紫檀透雕夔龙蝠纹架几案
### Red Sandalwood Side Table with Openwork Carving in Kui-dragon and Bat Design

**❶** 方几透雕蝠纹挡板细部
Detail of Openwork Carving in Bat Design on the Side Panel

**❷** 方几束腰
Detail of Frieze on the Square Parnel

**❸** 方几夔龙透雕纹饰挡板细部
Detail of Openwork Carving in Kui-dragon Design on the Side Panel

**❶ 牙板雕刻细部**
Detail of Caving on the Apron

## 紫檀带束腰西洋风格雕花长桌
### Red Sandalwood Long Table with Western Style Carved Apron

**❷ 牙板与桌腿雕刻细部**
Detail of Caving between the Apron and Leg

**❸ 牙板西洋风格纹饰雕刻**
Detail of Western Style Carving on the Apron

**❹ 内翻马蹄**
Detail of Varus Horseshoe

## 黑漆描金山水楼阁炕几
### Black Lacquer Bed Table with Painted Gold in Mountains, River, and Building Pattern

❷ 腿足内翻马蹄下踩圆珠
Detail of Varus Horseshoe with Round Metal Decoration

❸ 棍格牙头
Detail of the Apron

❹ 束腰雕菱形透空
Detail of Rhombic open Spaces of Carving on the Frieze

❶ 紫檀雕云纹顶竖柜局部
Detail of Red Sandalwood Compound Wardrobe with Carved Cloud Design

## 紫檀雕云纹顶竖柜
## Red Sandalwood Compound Wardrobe with Carved Cloud Design

❷ 柜门开光内雕刻山水人物 1
Detail of Carved Figures and Scenery on Door Panel 1

❸ 柜门开光内雕刻山水人物 2
Detail of Carved Figures and Scenery on Door Panel 2

❹ 柜门开光内雕刻山水人物 3
Detail of Carved Figures and Scenery on Door Panel 3

❶ 面叶
Detail of Hinge

## 硬木连三柜橱
## Hardwood Cabinet

❷ 合页
Detail of Hinge

❸ 牙条细部
Detail of the Apron

❹ 内翻马蹄
Detail of the Varus Horseshoe

❶ 古铜镜细部
Bronze Mirror

## 紫檀木边座嵌古铜镜大插屏
### Red Sandalwood Stand Table Screen with Bronze Mirror

❷ 屏心御制诗
Poem Authored by Emperor on the Central Panel of Screen

❸ 座架绦环板雕刻细部
Detail of Carving on the Panel of Stand

❹ 紫檀木墩子细部
Detail of Red Sandalwood Shoe Foot

## 紫檀木边座嵌玉人鸂鶒木山水图插屏
### Red Sandalwood Stand Table Screen with Jade and Chicken-wing Wood Inlay Depicting Figures and Scenery

❶ 站牙细部
Detail of Standing Spandrel

❷ 披水牙细部雕刻
Detail of Carved Slanted Apron

❸ 墩子细部
Detail of Shoe Foot

## 紫檀木雕花框极乐世界佛屏
## Pure Land Buddha Screen With Red Sandalwood Carved Flower Frames

❶ 绢绘细部
Detail of Painted Silk

❷ 绢绘佛像
Detail of Buddha Painted on the Silk

❸ 小佛龛内佛像
Detail of Buddha Statue in the Niche

须绘细部
Detail of Painted Silk

❶ 紫檀木雕屏风帽
Red Sandalwood Screen Top with Carving Decorations

## 紫檀木雕屏风
## Red Sandalwood Screen with Carving Decorations

❷ 屏风心题字
Detail of Inscription on the Screen

❸ 屏风木雕座 1
Detail of Wooden Table Screen Seat with Carving Decorations

❹ 屏风木雕座 2
Detail of Wooden Table Screen Seat with Carving Decorations

❶ 紫檀木边座雕鸂鶒木山水围屏雕云龙纹屏帽局部
Detail of Carved Dragon and Cloud Pattern on Screen Top Panel

## 紫檀木边座雕鸂鶒木山水围屏
Red Sandalwood Stand Folding Screen with Chicken-wing Wood Inlay Depicting Scenery

❷ 屏心山水纹雕刻
Detail of Carving on the Central Panel of Screen

❸ 须弥式屏座雕刻细部
Detail of Carving on Buddhist Pedestal

❹ 须弥式屏座圭角
Detail of Pedestal Foot

文渊阁宝座屏风陈设
Interior Display of Throne and Screen in Pavilion of Literary Profundity (Wenyuange)

❶ 雕云纹屏帽
Detail of Cloud Design
Carving on the Top

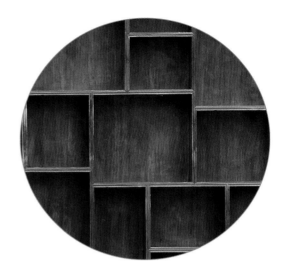

❷ 多宝槅细部
Detail of the Case

## 硬木多宝槅式围屏
## Hardwood Throne Surround Screen and Display Case

❸ 红漆描金细部 1
Detail of Red Lacquer and
Painted Gold 1

❹ 红漆描金细部 2
Detail of Red Lacquer and
Painted Gold 2

❺ 红漆描金细部 3
Detail of Red Lacquer and
Painted Gold 3

硬木多宝槅式围屏
Hardwood Throne Surround Screen and Display Case

① 木雕匾额
Horizontal Inscribed Board with Carved Frame

# 花梨木雕夔龙框绢地金字匾
## Horizontal Inscribed Board with Carved Frame

② 匾额雕刻细部 1
Detail of Carving on the Frame

③ 匾额雕刻细部 2
Detail of Carving on the Frame

④ 匾额雕刻细部 3
Detail of Carving on the Frame

漱芳斋内景陈设
Interior Scene of Lodge of
Fresh Fragrance (Shufangzhai)

漱芳斋内景陈设
Interior Scene of Lodge of Fresh Fragrance (Shufangzhai)

❶ 木雕细部 1
Detail of the Woodcarving 1

❷ 木雕细部 2
Detail of the Woodcarving 2

## 三友轩紫檀木透雕松竹梅窗
Red Sandalwood Window with Openwork Carving in Pine, Bamboo and Plum Blossom in Bower of Three Friends

❸ 木雕细部 3
Detail of the Woodcarving 3

❹ 木雕细部 4
Detail of the Woodcarving 4

❺ 木雕细部 5
Detail of the Woodcarving 5

大修前乾隆花园三友轩松竹梅窗户东侧
East Side of Window with Pine, Bamboo and Plum Blossom Decorations, Bower of Three Friends, Before Conservation

第二部分

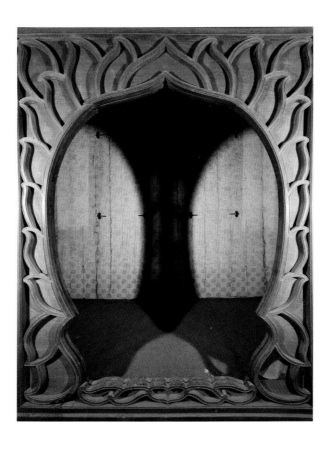

大修前乾隆花园养和精舍东进间木雕火焰纹圆光罩西侧面
West Side of Wooden Flame-Shaped Moon Gate in the East Bay, Supreme Chamber of Cultivating Harmony, Before Conservation

## 楠木雕莲花纹壶门式圆光罩
## Nanmu Pot-gate Style Aureola Corer Surrounded with Openwork Carving in Lotus Petal Pattern

❶ 紫檀贴雕夔龙绦环板裙板
Lower Section of the Partition with Carved Red Sandalwood Veneer in Kui-dragon Pattern

# 柏木灯笼框夹纱落地罩
# Cypress Partitions and Entablature with Inset Silk Gauze

❷ 湖绿色团龙八宝纹饰芝麻纱
Detail of Lake-green Silk Gauze in Coiled Dragon and Eight Treasures Pattern

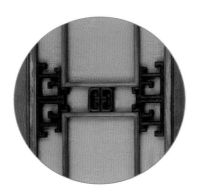

❸ 紫檀夔龙纹卡子花
Detail of Red Sandalwood Fillet in Kui-dragon Pattern

❹ 紫檀透雕夔龙纹花牙子
Detail of Red Sandalwood Bracket with Openwork Carving in Kui-dragon Pattern

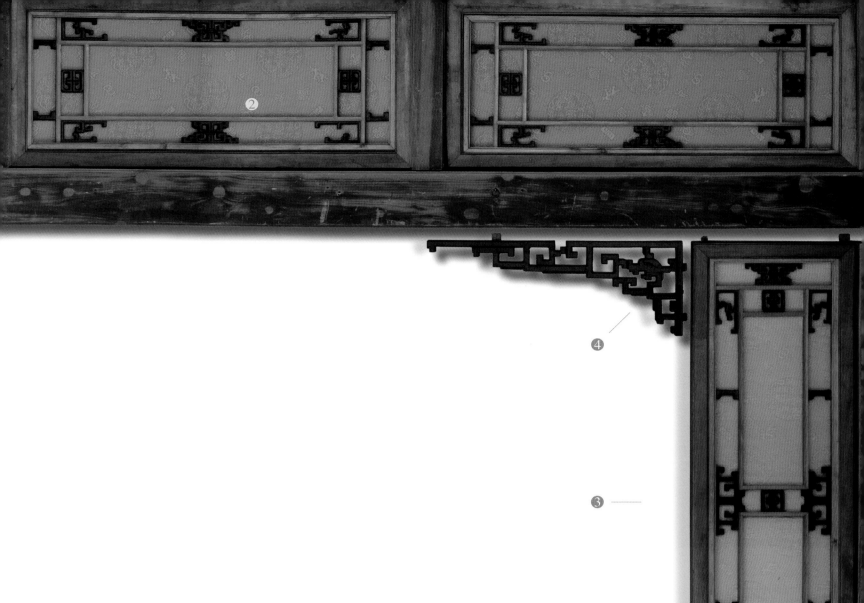

❶ 木雕博古隔扇裙板
Wood Screen Skirting Panel with Carved Decorations

## 紫檀木雕博古隔扇
## Red Sandalwood Carved Screen in Palace of Double Brilliance

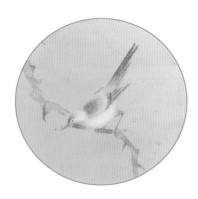  

❷ 隔扇嵌画
Detail of Screen with Painting Insets

❸ 木雕隔扇心
Detail of Carved Screen

❹ 木雕博古香炉
Detail of Carved Incense Burner Decoration

重华宫木雕隔扇
Carved Screen in Palace of Double Brilliance (Chonghuagong)

❶ 倦勤斋紫檀花梨木拼万字锦地贴雕竹黄花鸟图案炕檐板
Bed-sided Board with Swastika Patterned Background of Red Sandalwood and Padauk and Carved Flowers and Birds Veneer of Inner-bamboo-skin

## 倦勤斋紫檀木雕绳纹回纹嵌竹丝嵌玉落地罩
Red Sandalwood and Padauk: Rope-shape Rectangular Spiral Pattern Partitions with Bamboo Filament Marquetry and Jade Insets in Juanqinzhai (The Studio of Exhaustion from Diligent Service)

❷ 竹黄贴雕细部（类似）1
Detail of Carved Inner-bamboo-skin Veneer 1

❸ 竹黄贴雕细部（类似）2
Detail of Carved Inner-bamboo-skin Veneer 2

❹ 竹黄贴雕细部（类似）3
Detail of Carved Inner-bamboo-skin Veneer 3

141

倦勤斋东五间楼上紫檀花梨木拼万字锦地竹黄贴雕花鸟图槛墙木墙板
Partition Wall of the East Five Bays in Juanqinzhai( the Studio of Exhaustion from Diligent Service) with Swastika Patterned Background of Red Sandalwood and Padauk and Carved Inner-bamboo-skin Veneer of Flowers and Birds (Upper Storey)

倦勤斋东五间楼下贴雕竹黄山林百鹿图
Carved Inner-bamboo-skin Veneer of One Hundred Deer Motif in the East Five Bays of Juanqinzhai( the Studio of Exhaustion from Diligent Service)
(Ground Storey)

## 倦勤斋西四间斑竹纹彩绘隔扇窗及槛墙
Faux-bamboo Lattice Window and Wainscot in the West Four Bays of Juanqinzhai (the Studio of Exhaustion from Diligent Service)

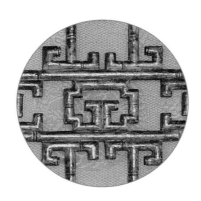

❶ 斑竹彩绘细部
Detail of Faux-bamboo Decoration

❷ 斑竹彩绘绦环板
Faux-bamboo Decorative Panel

❸ 斑竹彩绘细部
Detail of Faux-bamboo Decoration

Faux-bamboo Lattice Moon-shaped Gate

倦勤斋明殿西进间内景
Interior of West Room of Bright Hall of Juanqinzhai (Studio of Exhaustion from Diligent Service)

❶ 紫檀花梨木拼万字锦地
Swastika Patterned Background of Red Sandalwood and Padauk Wood

❷ 竹黄贴雕细部
Detail of Carved Inner-bamboo-skin Veneer

Bamboo Filament Marquetry

竹丝镶嵌

## 竹丝镶嵌玻璃博古橱
### Glass Antique Display Case with Bamboo Filament Marquetry Decoration

❶ 竹丝镶嵌细部 1
Detail of Bamboo Filament Marquetry Decoration 1

❷ 竹丝镶嵌细部 2
Detail of Bamboo Filament Marquetry Decoration 2

❸ 腿足细部
Detail of the Foot

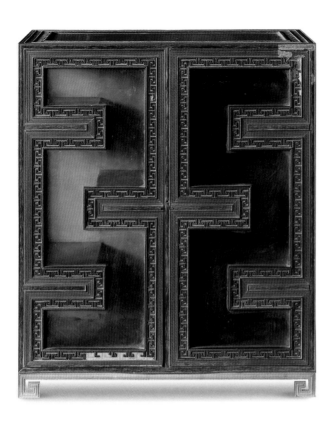

竹丝镶嵌玻璃博古橱
Antique Display Case with Bamboo Filament Marquetry Decoration

書齋迎愛日
兩字仰
天文居所勉
無倦臨民首
力勤新知期
受益舊學永
尊聞不息法
先憲孜孜念
惜分
己巳長至
月之下澣
御題

**❶ 竹丝镶嵌炕檐板**
Bed-sided Board with Bamboo Filament Marquetry Decoration

## 竹丝紫檀丝镶嵌万字锦地贴雕回纹嵌玉炕檐板
Bed-sided Board of Bamboo Filament and Red Sandalwood Filament Marquetry Background with Rectangular Spiral Pattern Carved Ebony Veneer and Jade Insets

**❷ 竹丝镶嵌细部**
Detail of Bamboo Filament Marquetry

**❸ 乌木回纹贴雕玉石镶嵌细部**
Detail of Rectangular Spiral Pattern Carved Ebony Veneer and Jade Insets

**❹ 玉石镶嵌细部**
Detail of Jade Insets

画珐琅镶嵌

Painted Enamel Inlay

① 掐丝珐琅镶嵌隔扇心细部
Detail of Cloisonne Decoration

符望阁紫檀回纹嵌掐丝珐琅灯笼框
夹纱臣工书画贴落隔扇心
Fuwangge Red Sandalwood Upper
Section of the Partition Screen with
Rectangular Spiral Shaped Cloisonne
and Inset Silk Gauze of Painting
and Calligraphy in the Parilion of
Expecting Good Omens

② 掐丝珐琅镶嵌隔扇心细部
Detail of Cloisonne Decoration

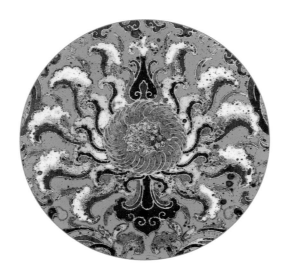

① 掐丝珐琅镶嵌细部 1
Detail of Cloisonne Inset 1

## 符望阁紫檀边框镶嵌掐丝珐琅炕檐板
## Fuwangge Red Sandalwood Frame Bed-sided Board with Cloisonne Insets

·图版·画珐琅镶嵌·

❷ 掐丝珐琅镶嵌细部 2
Detail of Cloisonne Inset 2

❸ 掐丝珐琅镶嵌细部
——菊花图案
Detail of Cloisonne Inset—
Chrysanthemum Pattern

❹ 掐丝珐琅镶嵌细部 3
Detail of Cloisonne Inset 3

❺ 掐丝珐琅镶嵌细部
——蝠纹图案 1
Detail of Cloisonne Inset
—Bat Pattern 1

❻ 掐丝珐琅镶嵌细部
——蝠纹图案 2
Detail of Cloisonne Inset
—Bat Pattern 2

❼ 掐丝珐琅镶嵌细部 4
Detail of Cloisonne Inset 4

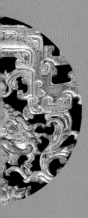

錾铜镶嵌

Chiseled Copper Inlay

紫檀木嵌铜鎏金缠枝纹落地罩
Red Sandalwood Frame Screen with Glazed Gold Inlay

❶ 錾铜鎏金装饰细部 1
Detail of Gilt and Copper Decoration 1

❷ 錾铜鎏金装饰细部 2
Detail of Gilt and Copper Decoration 2

## 符望阁紫檀边框錾铜鎏金装饰炕檐板
Fuwangge Red Sandalwood Frame Bed-sided Board with Gilt and Copper Decoration

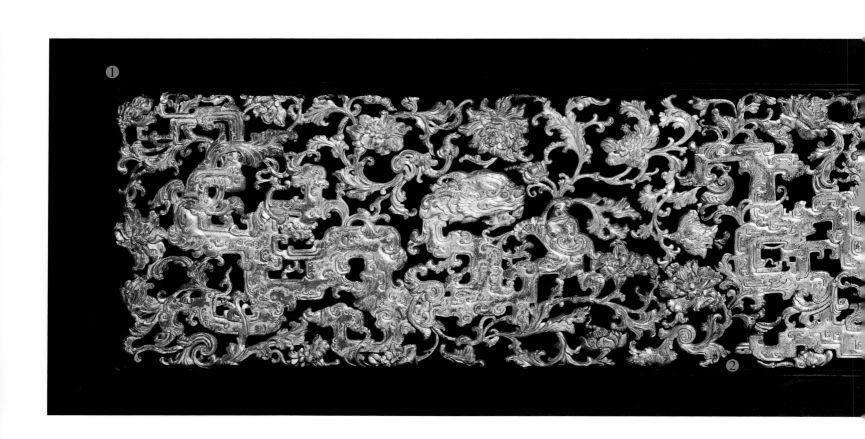

❸ 錾铜鎏金——夔龙图案
Detail of Gilt and Copper Decoration—Kui-dragon Design

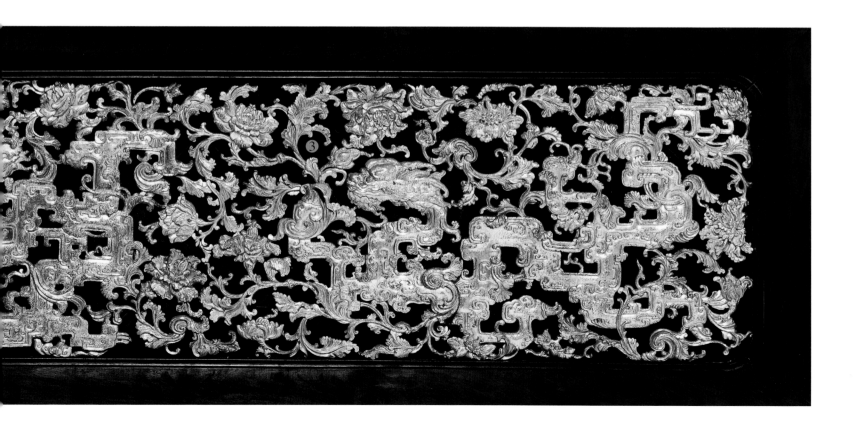

Mother-of-pearl Inlay

螺钿镶嵌

黑漆嵌螺钿像生博古槅[1]
Black Lacquer Simulated Display Cabinet with Mother-of-pearl Inlay

❶ 嵌螺钿装饰细部 1
Detail of Mother-of-pearl Inlay 1

注1：内有像生盆栽八件
Note1: Eight Simulated Plants Inside

❷ 嵌螺钿装饰细部 2
Detail of Mother-of-pearl Inlay 2

❶ 牙条嵌螺钿寿字纹饰 1
Detail of Mother-of-pearl Inlay of Character "Shou" (Longevity) Pattern 1

❷ 腿足细部
Detail of Foot

## 花梨木嵌螺钿长桌
## Padauk Wood Table with Mother-of-pearl Inlay

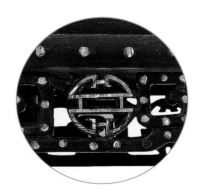

❸ 牙条嵌螺钿寿字纹饰 2
Detail of Mother-of-pearl Inlay of Character "Shou" (Longevity) Pattern 2

❹ 腿足细部 1
Detail of Foot 1

❺ 腿足细部 2
Detail of Foot 2

花梨木嵌螺钿长桌

Padauk Wood Table with Mother-of-pearl Inlay

长桌正面、侧面
Front and Side of the Wood Table

符望阁漆地嵌螺钿描金彩绘迎风板
Fuwangge Panel with Lacquer Background and Mother-of-pearl and Painted Gold Decorations

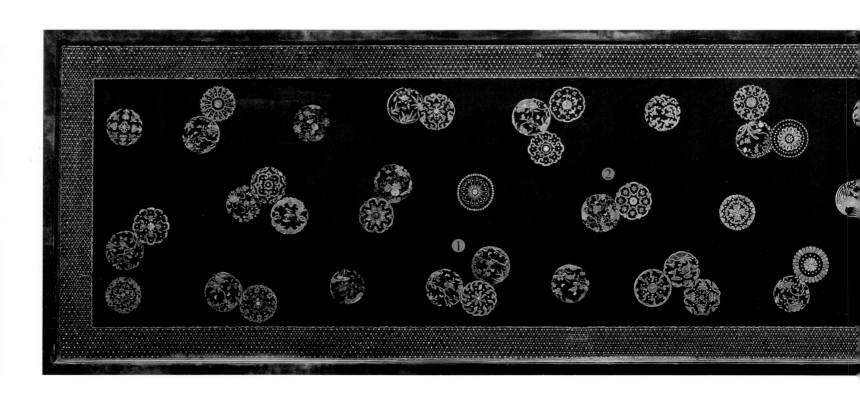

图版·螺钿镶嵌

❶ 漆地描金彩绘细部 1
Detail of Lacquer
Background and Painted
Gold Decorations 1

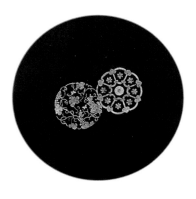

❷ 漆地描金彩绘细部 2
Detail of Lacquer
Background and Painted
Gold Decorations 2

❸ 漆地描金彩绘细部 3
Detail of Lacquer
Background and Painted
Gold Decorations 3

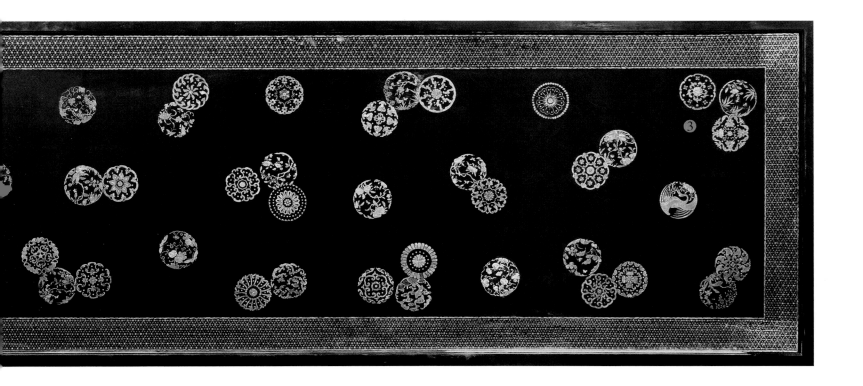

**❶ 玻璃画细部**
Detail of Glass Painting

## 萃赏楼紫檀拼竹嵌螺钿玻璃画隔心黑漆描金花卉绦环板裙板隔扇

Red Sandalwood Partition Screens with Upper Sections of Bamboo, Mother-of-pearl, Glass Painting Decorations and Lower Sections of Black Lacquered Panel with Gilding Decorations in the Building for Enjoying lush Scenery(Cuishanglou)

❷ 拼竹嵌螺钿隔扇心细部
Detail of Mother-of-pearl and Bamboo Decoration

❸ 黑漆描金裙板细部
Detail of Black Lacquered Panel with Gilding

❹ 嵌螺钿细部
Detail of Carved Mother-of-pearl Decoration

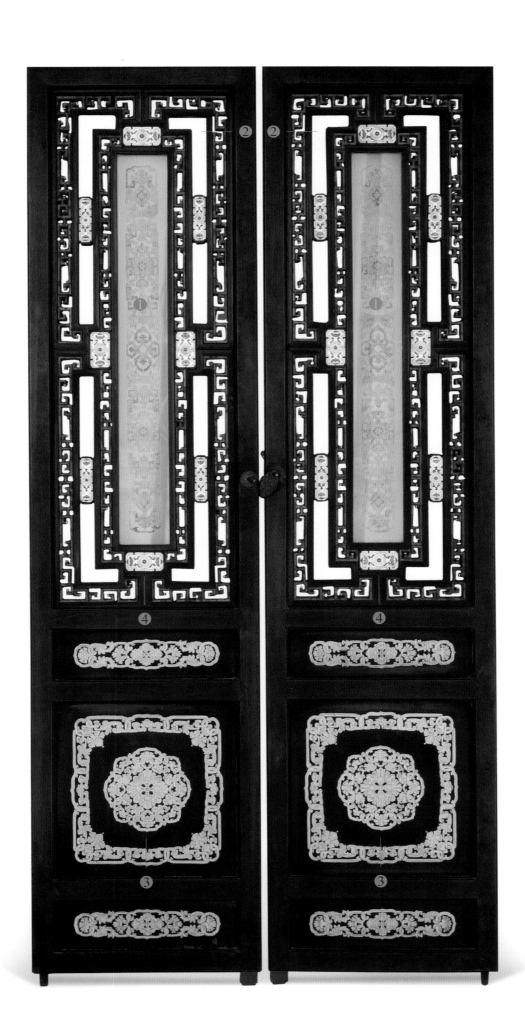

Jade, Precious Stone
and Marble Inlay

玉石、宝石、大理石镶嵌

## 紫檀嵌玉绳纹拱璧炕几
### Red Sandalwood Bed Table with Jade Inlay and Funiform Decorations

❶ 嵌玉绳纹拱璧牙子
Detail of the Apron with Jade Inlay and Funiform Decorations

❷ 嵌玉绳纹拱璧牙子
Detail of the Apron with Jade Inlay and Funiform Decorations

❸ 内翻回纹马蹄
Detail of Varus Fretted Horseshoe

## 紫檀木嵌玉松石玉竹梅玻璃插屏
### Red Sandalwood Table Screen with Jade Pine, Bamboo and Plum Blossom Inlay on Glass

❶ 翠玉雕松枝图案
Detail of Green Jadeite Pine Branches Inlay

❷ 白玉雕梅花图案
Detail of White Jade Plum Blossom Inlay

❸ 灵芝图案细部
Detail of Ganoderma Lucidum Pattern

· 图版 · 玉石、宝石、大理石镶嵌 ·

· 第二部分 ·

**❶ 插屏嵌玉人细部**
Detail of Jade Figure Inlay

## 紫檀边鸡翅木嵌玉人插屏
## Red Sandalwood Frame Table Screen with Carved Chicken-wing Wood Decorations and Jade Figure Inlays

**❷ 插屏细部 1**　　　　　　　　　　**❸ 插屏细部 2**　　　　　　　　　　**❹ 插屏细部 3**
Detail of Table Screen 1　　　　　　Detail of Table Screen 2　　　　　　Detail of Table Screen 3

## 紫檀木嵌玉雕松竹梅盆架
### Red Sandalwood Stand with Jade Pine, Bamboo and Plum Blossom Inlay

❶ 翠玉雕松枝图案
Detail of Green Jadeite Pine Branches Inlay

❷ 白玉雕梅花图案
Detail of White Jade Plum Blossom Inlay

❸ 翠玉雕竹叶图案
Detail of Green Jadeite Bamboo Leaves Inlay

❹ 翠玉雕柏枝图案
Detail of Green Jadeite Cypress Branches Inlay

❶

❷

❶ 波浪纹铜框嵌玉松枝透雕
Detail of Copper Frame with Jadeite Pine Branches Inlay

❷ 波浪纹铜框嵌玉梅花透雕
Detail of Copper Frame with White Jade Plum Blossoms Inlay

❸

❹

❸ 波浪纹铜框嵌玉竹叶透雕
Detail of Wave Patterned Copper Frame with Jadeite Bamboo Leaves Inlay

❹ 波浪纹铜框嵌玉柏枝透雕
Detail of Wave Patterned Copper Frame with Jadeite Pine Branches Inlay

## 雕天然木框嵌玉石花鸟纹挂屏
## Hanging Screen with Inlaid Stones Depicting Bird-and-flower Scenes

❶ 嵌玉花草图案细部 1
Detail of Flower and Grass Design Inlay 1

❷ 嵌玉花鸟图案细部
Detail of Bird-and-flower Design Inlay

❸ 嵌玉花草图案细部 2
Detail of Flower and Grass Design Inlay 2

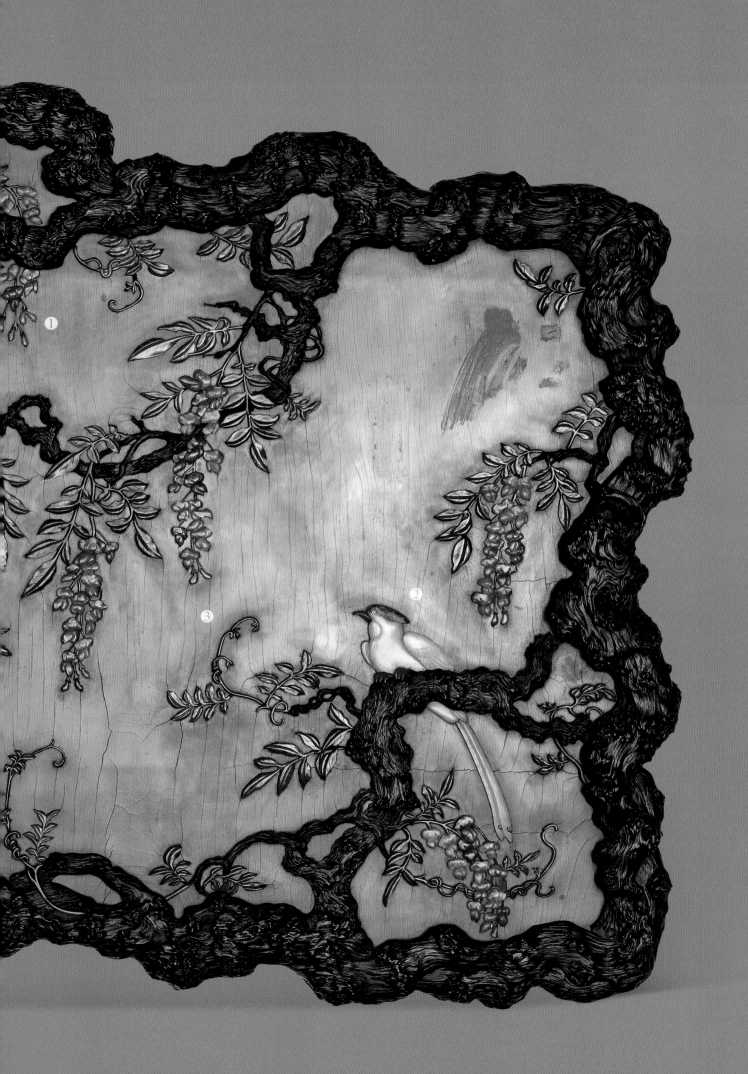

大修前乾隆花园符望阁首层西次间中进间西缝落地罩西侧
West Side of a Screen on the First Floor of Fuwangge,
Qianlong Garden, Before Conservation

图版·玉石、宝石、大理石镶嵌·

## 符望阁紫檀框嵌玉迎风板
## Fuwangge Red Sandalwood Frame Panel with Jade Inlay

图版·玉石、宝石、大理石镶嵌·

❶ 嵌玉竹叶图案
Detail of Jade Bamboo Leaves Inlay

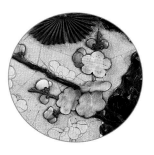

❷ 嵌玉梅花图案 1
Detail of Jade Plum Blossoms Inlay 1

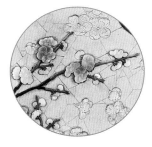

❸ 嵌玉梅花图案 2
Detail of Jade Plum Blossoms Inlay 2

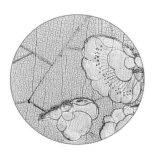

❹ 米色漆地冰梅纹
Detail of Beige Lacquer Background with Cracked-ice and Plum Blossom Pattern

❺ 嵌象骨松枝图案
Detail of Bone Pine Branches Inlay

❻ 嵌蜜蜡松塔图案
Detail of Copal Resin Pinecone Inlay

191

## 符望阁紫檀嵌玉沉香木雕炕檐板
### Fuwangge Red Sandalwood Bed-sided Board with Jade Insets and Agalloch Carving

图版·玉石、宝石、大理石镶嵌·

❶ 嵌玉沉香木雕桃树细部 1
Detail of Jade Insets and
Agalloch Carving 1

❷ 沉香木雕细部
Detail of Agalloch Carving

❸ 嵌玉沉香木雕桃树细部 2
Detail of Jade Insets and
Agalloch Carving 2

## 三友轩紫檀嵌玉卷草花卉槛窗
## Red Sandalwood Window with Jade Insets and Scrolled Grass and Flower Pattern Decoration in Bower of Three Friends

❶ 嵌玉槅心
Detail of Jade Insets

❷ 书画贴落
Detail of Inset Silk Gauze of Calligraphy

❸ 紫檀贴雕万字不到头图案绦环板细部
Detail of Panel with Red Sandalwood Carved Veneer Decoration

## 符望阁嵌玉沉香木雕槛窗
Fuwangge Window with Jade Insets and Agalloch Carving

**❶** 嵌玉沉香木雕细部 1
Detail of Jade Insets and Agalloch Carving 1

**❷** 嵌玉沉香木雕细部 2
Detail of Jade Insets and Agalloch Carving 2

**❸** 嵌玉沉香木雕细部 3
Detail of Jade Insets and Agalloch Carving 3

围屏边框及下裙板用紫檀木制作，屏分为16扇。屏心落堂作，髹黑漆地，用"周制凸嵌法"以0.5厘米厚的白玉片镶嵌成佛教中十六罗汉像。显现出剪纸或贴画的艺术效果。加上黑地和白玉间的色彩反差，使画面形象更为突出。十六罗汉不仅姿态各异，其神情表情也各不相同，或庞眉大目、朵颐隆鼻，或皱额凹头、深目翻鼻，极具胡貌梵相，生动传神。每个罗汉的上方均嵌有罗汉的名称和乾隆皇帝所做的罗汉赞。裙板呈正方形，分别雕刻四条拐子龙纹。

The screen is comprised of sixteen sets of panels and their frames and lower sections are made of red sandalwood. Inlaid in the wooden frame, the plaques are decorated with black lacquer background and white jade arhat insets with the thickness of 0.5 cm. The contrast between the black background and white insets highlights the figures of the arhats, which hold various postures and lively expressions. Above each arhat figure, the area is inlaid with the name of the arhat and the inscriptions written by Qianlong Emperor. The lower sections of the screen are the skirting panels with carved spiral dragon pattern.

## 硬木嵌玉十六罗汉像屏
## Wooden Screen with Inlaid Jade Plaques Portraying the Sixteen Arhats

❶ 嵌玉罗汉像细部1
Detail of Arhats Portrait With Jade Inlay 1

❷ 嵌玉罗汉像细部2
Detail of Arhats Portrait With Jade Inlay 2

❸ 嵌玉赞词细部3
Detail of Inscription with Jade Inlay 3

·图版·玉石、宝石、大理石镶嵌·

·第二部分·

· 图版 · 玉石、宝石、大理石镶嵌 ·

· 第二部分 ·

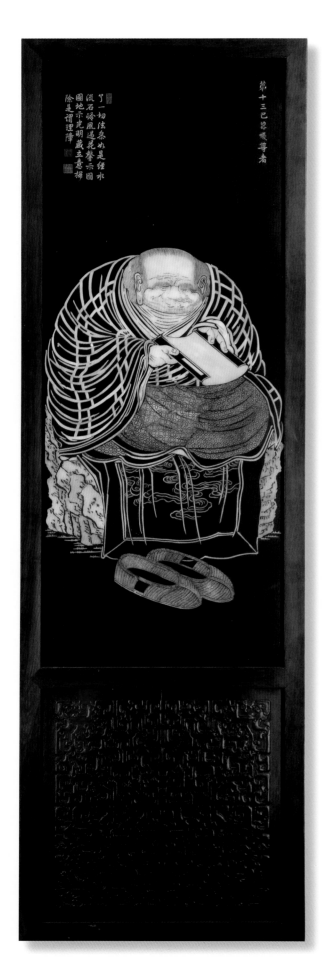

**❺ 紫檀雕刻裙板（屏风）**
Red Sandalwood Skirting Panel with Carved Decorations (Screen)

屏风背面，屏心黑漆地，以金漆彩绘竹子、牡丹、藤萝、荷花、桃花、月季、桂花、美人蕉、茶花、水仙、菊花、秋葵等各式折枝花卉，至今色彩鲜艳明亮。裙板与正面相同，屏风原陈清宫乾隆花园云光楼大佛龛的莲花台上，为三面直角陈设，正面十扇，两侧各三扇。

The plaques on the back of the screens are black lacquer background, decorated with gold painted bamboo, poeny, Chinese wisteria, lotus, peach blossom, Chinese rose, sweet olive, cannalily, camellia, narcissus, chrysanthemum and okra. The screen was originally displayed on the lotus platform of a niche in the Building of Luminous Cloud in the Qianlong Garden. Ten sets were placed in the middle with both sides displaying three sets vertically.

  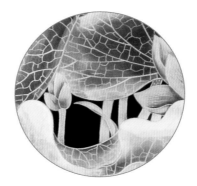 

**❶~❹ 黑漆描金荷花图案细部**
Black Lacquer Background Painted with Golden Lotus Patterns

· 图版 · 玉石、宝石、大理石镶嵌 ·

· 第二部分 ·

・图版・玉石、宝石、大理石镶嵌・

・第二部分・

·图版·玉石、宝石、大理石镶嵌·

·第二部分·

・图版・玉石、宝石、大理石镶嵌・

・第二部分・

·图版·玉石、宝石、大理石镶嵌·

·第二部分·

・图版・玉石、宝石、大理石镶嵌・

・第二部分・

图版・玉石、宝石、大理石镶嵌・

符望阁紫檀边框百宝嵌炕檐板
Fuwangge Red Sandalwood Frame Bed-sided Board with Multi-precious Objects Inlay

·图版·玉石、宝石、大理石镶嵌·

❶~❻ 百宝嵌图案细部
Detail of Multi-precious Objects Inlay

·图版·玉石、宝石、大理石镶嵌·

·第二部分·

·图版·玉石、宝石、大理石镶嵌·

·第二部分·

❶ 玉石镶嵌细部
Detail of Stone Inlay

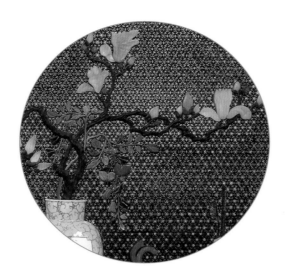

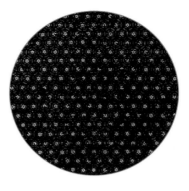

❷ 螺钿镶嵌细部
Detail of Mother-of-pearl Decorated Background

## 符望阁紫檀木框嵌螺钿博古百宝嵌迎风板
Fuwangge Red Sandalwood Frame Panel with Stone Inlays and Mother-of-pearl Decorated Background

图版·螺钿镶嵌

❸ 紫檀木框嵌螺钿博古百宝嵌迎风板细部
Detail of the Red Sandalwood Frame Panel with Stone Inlay and Mother-of-pearl Decorated Background

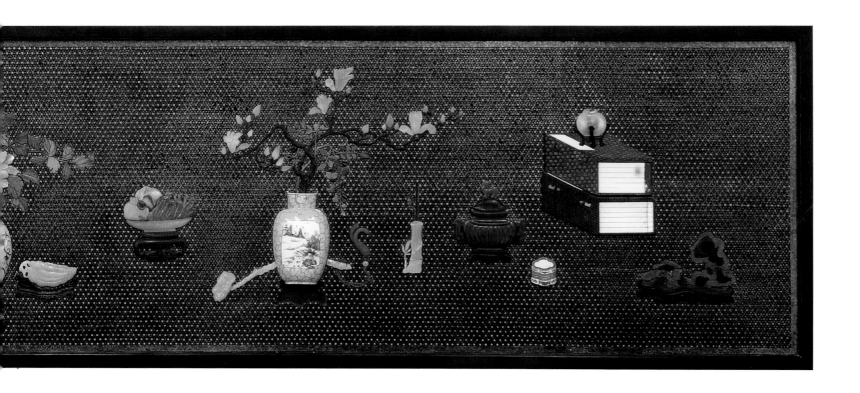

**❶ 镂雕嵌玉围子**
Engraved Seat Rail with Jade Inlay

## 黄花梨镂雕嵌玉围子罗汉床
## Padauk Wood Arhat Bed with Openwork Carved Seat Rail and Jade Inlay

**❸ 肩部云纹雕刻**
Detail of Carved Cloud Pattern

**❹ 牙板拐子纹雕刻细部**
Detail of Rectangular Spiral Carving on the Apron

**❺ 内翻马蹄床脚**
Varus Horseshoe Foot

❷ 罗汉床侧面
Side of Arhat Bed

黄花梨镂雕嵌玉围子罗汉床
Padauk Wood Arhat Bed with Openwork Carved Seat Rail and Jade Inlay

**❶ 椅背细部**
Detail of the Backrest

## 硬木嵌石七屏式罗汉床
## Hardwood Fire-screen Arhat Bed with Marble Insets

 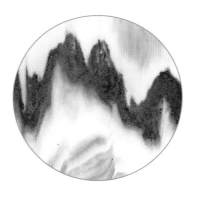 

**❸ 靠背盘长图案雕刻**
Detail of Carving on the Backrest

**❹ 靠背镶天然大理石细部**
Detail of Natural Marble Insets

**❺ 三弯腿细部**
Detail of Three Curved Leg

图版・玉石、宝石、大理石镶嵌

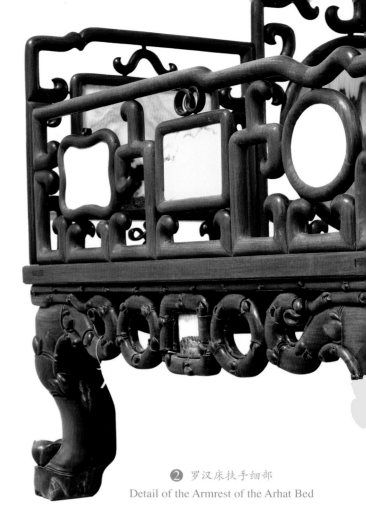

② 罗汉床扶手细部
Detail of the Armrest of the Arhat Bed

第二部分

・图版・玉石、宝石、大理石镶嵌・

・第二部分・

图版・玉石、宝石、大理石镶嵌・

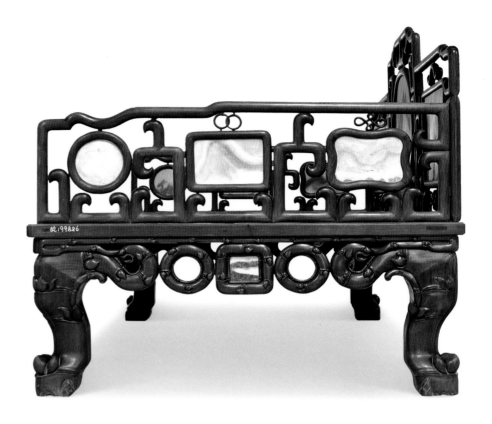

第二部分

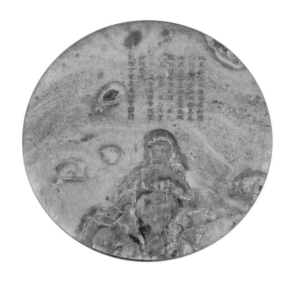

**❶ 屏心天然石料观音像**
Detail of Stone Inset of Natural Image of Kwan-Yin

## 紫檀嵌天然石观音菩萨像插屏
## Red Sandalwood Table Screen Inset with Natural Stone Kwan-Yin Boddhisattva's Image

**❷ 屏心镌刻御制赞辞**
Detail of Stone Inset Incised with Inscriptions by Qianlong Emperor

**❸ 观音像细部**
Detail of Kwan-Yin

**❹ 屏心周围阴刻佛经细部**
Detail of Red Sandalwood Frame Incised with Buddhist Sutras

## 紫檀木雕花座嵌珐琅西洋人物插屏
### Red Sandalwood Frame Table Screen with Cloisonne Inset of Western Figures

❶ 屏心画珐琅图案细部
Detail of Cloisonne Inset

❷ 紫檀批水牙细部
Detail of Red Sandalwood Base

❸ 紫檀绦环板细部
Detail of Red Sandalwood Web Plate

·图版·玉石、宝石、大理石镶嵌·

·第二部分·

## 紫檀木框玻璃画三清图插屏
## Red Sandalwood Table Screen with taoist Glass Painting

❶ 玻璃画梅花图案  
Detail of Plum Blossom Glass Painting

❷ 绦环板上浮雕刻拐子纹  
Detail of Carved Rectangular Spiral Pattern

❸ 卷云腿足  
Detail of Foot with Scrolled Cloud Design

・图版・玉石、宝石、大理石镶嵌・

・第二部分・

① 隔扇心嵌粉彩瓷片局部
Detail of Porcelain Inlay

## 延趣楼紫檀雕回纹嵌瓷片夹纱槛窗
Red Sandalwood Partitions Window with Rectangular Sprial Pattern, Porcelain Inlay and Silk Gauze Inset in Building of Extenging Delight

② 隔扇心西番莲粉彩瓷片
Detail of Porcelain Inlay

③ 嵌瓷片雕回纹隔扇心
Detail of Carved Rectangular Sprial Pattern and Porcelain Decoration

④ 绦环板嵌青花矾红彩瓷片
Detail of Tile in Iron-red Enamel and Undergalzed Blue

嵌青花瓷片夹纱双面绣方窗
Square Window with Double-sided Silk Gauze and Blue-white Porcelain Inlay

隔扇心嵌粉彩瓷片局部
Detail of Porcelain Inlay

Lacquer Carving Technique

雕漆工艺

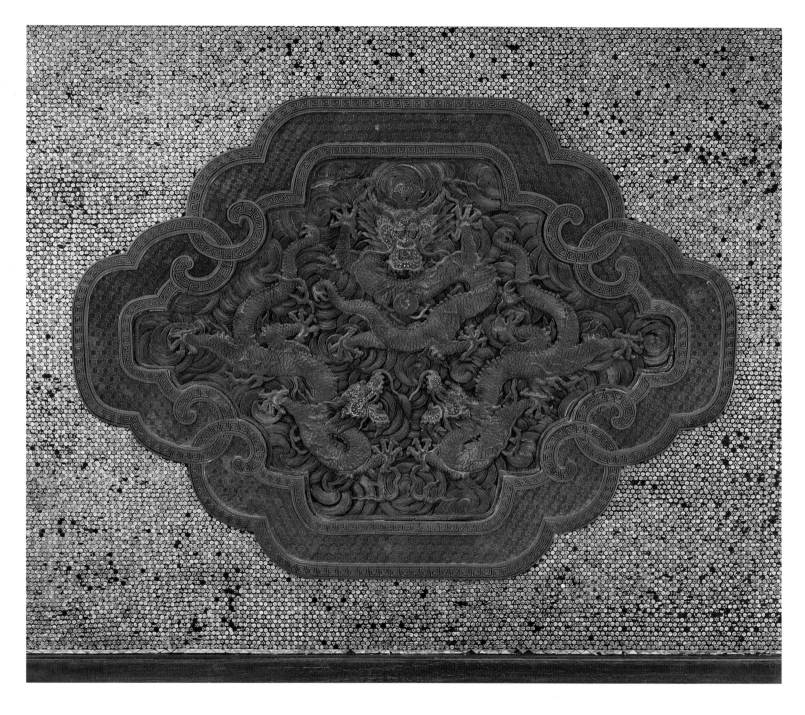

符望阁紫檀木框嵌螺钿雕漆龙纹迎风板细部
Detail of Fuwangge Red Sandalwood Frame Panel with Caved Lacquer Decorations, Dragon Patterns and Mother-of-pearl Decorated Background

## 符望阁紫檀木框嵌螺钿龙纹雕漆迎风板
Fuwangge Red Sandalwood Frame Panel with Caved Lacquer Decorations, Dragon Patterns and Mother-of-pearl Decorated Background

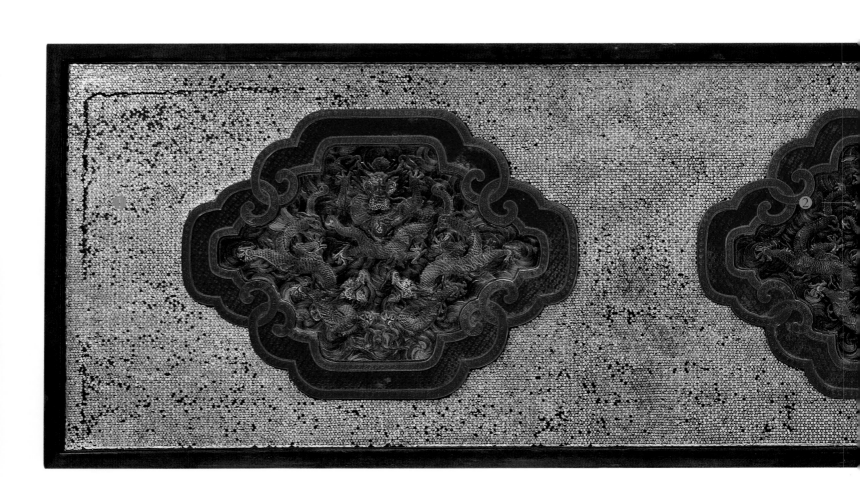

① 雕漆龙纹迎风板细部 1
Detail of Carved Lacquer in Dragon Pattern 1

② 雕漆龙纹迎风板细部 2
Detail of Carved Lacquer in Dragon Pattern 2

③ 雕漆龙纹迎风板细部 3
Detail of Carved Lacquer in Dragon Pattern 3

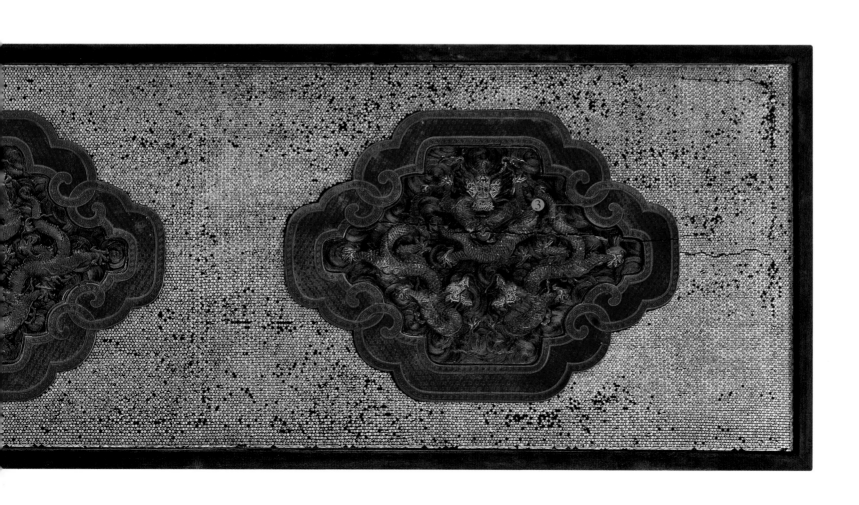

Various Materials and Techniques

多种材质及其工艺

**❶ 搭脑竹丝镶嵌细部**
Detail of Frame Backrest with Bamboo Filament Marquetry

## 攒竹嵌玉石七屏式宝座
（竹丝镶嵌部分）
Seven-screen Throne with Gathering Bamboo and Precious Stone Inlay
(Bamboo Filament Marquetry)

**❷ 靠背竹丝镶嵌边框**
Detail of Frame with Bamboo Filament Marquetry

**❸ 竹丝镶嵌细部**
Detail of Bamboo Filament Marquetry

**❹ 牙板竹丝镶嵌细部**
Detail of Connecting Plates with Bamboo Filament Marquetry

**❶ 靠背嵌玉石花卉图案**
Backrest with Flower Pattern with Precious Jade Inlay

## 攒竹嵌玉石七屏式宝座
（玉石镶嵌部分）
Seven-screen Throne with Gathering Bamboo and Precious Stone Inlay
（Precious Jade Inlay）

**❷ 座面侧沿嵌玉石装饰**
Detail of Side of the Throne with Jade Inlay Decoration

**❸ 嵌玉石花卉图案细部 1**
Detail of Flower Pattern with Precious Jade Inlay 1

**❹ 嵌玉石花卉图案细部 2**
Detail of Flower Pattern with Precious Jade Inlay 2

❶ 靠背背面漆地描金花卉图案
Back Side of Lacquer Painted with Gold in Flower Pattern

## 攒竹嵌玉石七屏式宝座
（漆地描金部分）
Seven-sereen Throne with Gathering Bamboo and Precious Jade Inlay
（Lacquer Painted with Gold）

❷ 漆地描金花卉图案细部1
Detail of Lacquer Background Painted with Gold in Flower Pattern 1

❸ 漆地描金花卉图案细部2
Detail of Lacquer Background Painted with Gold in Flower Pattern 2

❹ 漆地描金蝠纹细部
Detail of Lacquer Background Painted with Gold in Bat Pattern

❶~❷ 须弥座束腰玉石镶嵌花卉图案
Detail of Precious Jade Inlay of Frieie with Flower Pattern

## 攒竹嵌玉石缂丝围屏
（正面）
Chinese Silk Tapesty Throne Folding Screen with Precious
Jade Inlay and Bamboo Filament Marquetry
（Front）

①

侧屏
Side Section of the Screen

屏帽玉石镶嵌花卉图案
Detail of Precious Jade Inlay on the Top of the Screen with Flower Pattern

站牙竹丝镶嵌拐子纹
Detail of Bamboo Filament Marquetry in Rectangular Spiral Pattern

屏心四周竹丝镶嵌万字不到头图案
Detail of Bamboo Filament Marquetry in Swastika Pattern

须弥座束腰
Detail of Precious Jade Inlay on the Base of the Screen

·图版·多种材质及其工艺·

· 图版 · 多种材质及其工艺 ·

屏帽玉石镶嵌花卉图案
Detail of Precious Jade Inlay on the Top of the Screen

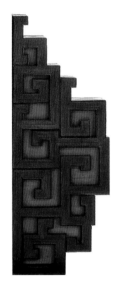

站牙竹丝镶嵌拐子纹
Detail of Bamboo Filament Marquetry in Rectangular Spiral Pattern

屏心边框竹丝镶嵌夔龙纹
Detail of Bamboo Filament Marquetry in Kui-dragon Pattern

须弥座托腰竹丝镶嵌"几马达"
Detail of Bamboo Filament Marquetry on the Base of the Screen

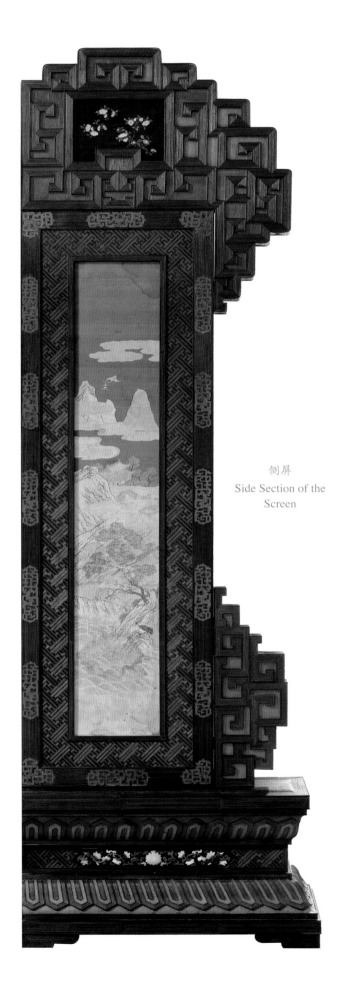

侧屏
Side Section of the Screen

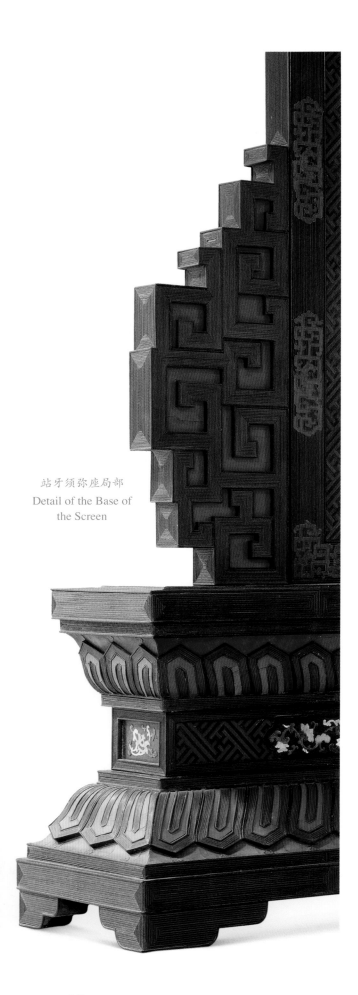

站牙须弥座局部
Detail of the Base of the Screen

竹丝镶嵌细部
Detail of Bamboo Filament Marquetry

须弥座束腰侧面玉石镶嵌花卉图案
Detail of Precious Jade Inlay on the Base of the Screen

竹丝镶嵌细部
Detail of Bamboo Filament Marquetry

左：侧屏心缂丝海屋添筹图
Left: Detail of Chinese Silk
Tapestry Inset of the Side Section

右：侧屏心缂丝海屋添筹图
Right: Detail of Chinese Silk
Tapestry Inset of the Side Section

下页：正中屏心缂丝海屋添筹图
The next page: Detail of Chinese
Silk Tapestry Inset of the Central
Section

❶ 背面屏心黑漆描金
Back of Black Lacquer Background Painted with Gold

# 攒竹嵌玉石缂丝围屏
（背面）
Chinese Silk Tapestry Throne Folding Screen with Precious Jade Inlay and Bamboo Filament Marquetry
(Back)

❷ 围屏背面须弥座束腰黑漆描金缠枝花卉图案
Detail of Black Lacquer Background Painted with Gold in Intertwined Flower Pattern on the Frieze of the Back of the Folding Screen

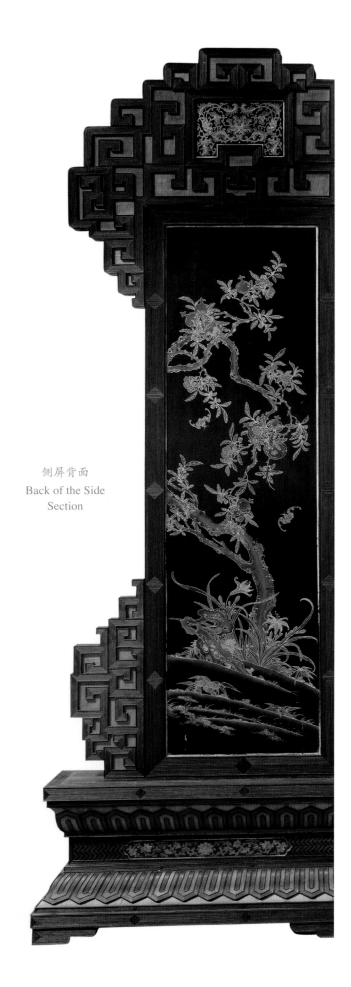

侧屏背面
Back of the Side Section

眉板黑漆描金
Detail of Black Lacquer on the Back of the Folding Screen

攒竹黑漆描金屏帽
Detail of Bamboo Filament Marquetry and Black Lacquer Background Painted with Gold Decorations on the Back

竹丝镶嵌细部
Detail of Bamboo Filament Marquetry

左、右及下页：
围屏背面屏心黑漆描金
Detail of Black Lacquer
Background Painted with Gold on
the Back of the Folding Screen

270

# Chapter III Appendix

第三部分 · 附录

# Restoration Technique on Hardwood Furniture

## 硬木家具装饰修缮技艺

古典家具木器修复把传统文化、传统工艺与现代科学技术结合起来，是一项技术性、艺术性很强的工作，古典家具木器的类别不同，修复方法也不同，因此要综合掌握多方面的知识，结合科学的措施灵活运用。

### 自然因素影响

导致古典家具装修损坏的自然因素主要包括温湿度、紫外线、红外线、有害气体、霉菌、害虫等方面。

当室温超过30℃、相对湿度高于70%时为高温高湿，这时易滋生霉菌、害虫等，并使其

Classic furniture conservation and restoration is a technical and artistic work, which combines the traditional culture, crafts, and modern technology. The restoration method varies from one to another while the furniture and materials are different; therefore, comprehensive knowledge is required, and scientific measures are flexibly applied as well.

### Natural factors

Natural factors including temperature, humidity, ultraviolet and infrared rays, toxic gas, molds, insects, etc. deteriorate hardwood furniture decorations.

Usually, when temperature goes above 30℃, and the relative humidity above 70% inside buildings, the molds and insects incline to breed fast, as a result, furniture will be eroded and deteriorated, and various chemical reactions are caused, even developed quickly under such high temperature and humidity condition. Besides, variance of temperature and humidity leads to the expansion and shrinkage of the

生长繁殖速度加快，加速了对古典家具的侵蚀及污染，各种化学反应速度也会因高温高湿环境而加剧，使各种腐蚀过程加快。此外，温湿度的变化会引起木材的湿胀干缩现象，导致木质古典家具木器干裂或变形。太阳光对木器有损害作用，其中红外线使木器表面温度升高、湿度下降，造成翘曲和脆裂；紫外线能使漆膜受到破坏而退色、脱落，同时紫外线还破坏木纤维结构，降低机械强度。

损坏古典家具木器的有害气体主要包括臭氧（$O_3$），二氧化硫（$SO_2$）、硫化氢（$H_2S$）、氮氧化物（$NO_x$）、氯气（$Cl_2$）等。

由于木材是天然高分子复合材料，硬度适中，营养丰富，因此，木质家具装修在很容易遭到各种真菌、昆虫的危害。引起木材缺陷损坏的真菌主要有木腐菌、软腐菌、霉菌或细菌。危害木器木材的昆虫主要有长蠹、粉蠹及某些种类的天牛、白蚁等。

## 修复原则

首先根据器物的类型确定修复原则。如文化价值或历史价值较高的家具装修，要严格遵循"按原样修复"和"修旧如旧"的原则修复；对于流传于民间的，采用比较名贵材料制作的古典家具，或者传世数量较大的普通柴木家具，可以根

wood of the furniture and even crack and deformation. Furthermore, the sunlight harms the furniture; the infrared rays cause the surface temperature going up and humidity going down, then curves and cracks will appear. While the ultraviolet ray damages the lacquer layer of the furniture, color fading and lacquer falling are caused, and meanwhile, the ultraviolet ray damages the fabric structure of the wood and decreases mechanical strength of the wood. The toxic gases, which damage the classical furniture, include ozone, sulfur dioxide, hydrogen sulfide, nitrogen oxide, chlorine, etc.

The natural wood, which has moderate hardness and plentiful nutrition, is composed of polymer composite. It is often damaged by bacteria such as wood-rotting fungi, soft-rot fungi, and insects like moth, longicorn, white ant, etc.

## Restoration principles

The types of the objects determine the restoration principles. The furniture with high cultural and/or historical value will be conserved strictly on the basis of the principles of "restoring to its original shape" and "restoring to its original state". The folk classic furniture made of precious wood, or classic furniture with large amount made of common wood sometimes are restored differently according to their condition, either "keep its original shape" or "keep its original state", that is, to restore in a simple way in order to preserve the surface of the color and grain of the wood, in some cases, people even work over the old furniture.

The conservators must have a good knowledge of the age of the furniture, characteristic of the materials, the mortise and tenon joints, coating and painting, etc. while conserving the classic wood furniture. Therefore, the same materials, characteristics, manufacturing techniques and structures can be preserved best. Meanwhile, the conservators must obey the principles of conservation reversibility. That is to say, one should reverse the objects to its original condition when any measures taken failed.

据不同情况，采取"保旧"、"留皮"（即古典家具脱漆后保留花皮做简单处理，保留老家具的色调和木质效果）或"翻新"处理。

总之，修复古典家具木器，应对所修家具木器的时代背景、材料性能、榫卯结构、髹饰工艺等十分熟悉，尽量保持相同的材料、形式特征、制作手法、构造特点等。同时要遵循古物修复的可逆性原则，即修复失败时应该能够恢复到原状。这就要求在修复时不用或尽量少用铁钉和化学粘合剂，以免破坏古典家具木器榫卯接合、易于拆修的特点。

In this case, nails and chemical adhesive should be used as less as possible in the conservation to avoid the damages to the mortise and tenon joints.

## Restoration tools

### Woodworking tools

Traditional woodworking tools are divided into rulers, saws, planes, chisels, etc. Rulers are used for measurement, they are folding ruler, zigzag rule, triangle rule, ink marker, line-marking knife, compasses, etc. Frame saws and knife saws are the most common saws in woodworking. According to their size and function planes are divided into long plane, medium plane, short plane, straight plane, line plane and so on. Chisels are used for making holes and carving, they are various in shape, such as flat, round, oblateness, and oblique. Besides, carpenters often have drills, hammers, axes, wood rasps, etc.

## 修复工具

### 1. 木工工具

传统木工工具分为测量工具、锯切工具、刨削工具、凿眼工具等。常用的测量工具有折尺、曲尺、三角尺、墨斗、划线刀、圆规等。木工常用的锯有框架锯和刀锯。刨子按刨身的长短及其用途，可分为长刨、中刨、短刨、线刨、蜈蚣刨等。开榫眼及雕刻都要用到各种形式的凿子，常用的有平凿、圆凿、扁凿、斜等。另外，木工制作时常用的还有钻、斧子、锤子、木锉等工具。

### 2. 髹饰工具

主要有漆刮、漆刷、罗筒、帚笔等。漆刮主要用于调刮漆克、调彩漆、取漆等，分为大号、中号和小号。漆刷是涂漆专用刷子，其材料有马尾、牛尾、猪鬃、人发等，以人发制作的质量较好。罗筒主要用于撒粉，一般用鹅毛管制作而成，将两端削成斜口，一端蒙上细绢，另一端装入金银粉，集中使用。帚笔即"扫笔"，用羊毫制作，大的用来扫除灰尘，小的扫除金银细粉。

### Lacquer-painting tools

Lacquer-painting tools include scraper, brush, pipe, broom brush, and so on. Scrapers are used for scraping, mixing color, taking lacquer. They are big, medium and small in size. Brushes are used for pasting up lacquer. They are made of the fur of horsetails land oxtails, big bristles, hairs, etc. The best scraper is made of human hair. Pipe is used for scattering powder. It is usually taken at the end of goose feather, both ends are cut into oblique, one end is covered by fine silk, the other is filled with the golden or silver powders. The broom brush is made of goat hairs, the bigger one can clean dust, and smaller one can sweep golden and silver powders.

## Restoration technique sequence

### Preparation

a) Documentation: photograph the furniture firstly, give the description, and record the damaged parts and the extent of the damages.

b) Restoration planning: according to the extent of the damages make a completed and detailed plan for restoration, formulate the related technical criteria, fill the technical process table.

c) Dismantling: observe the furniture in great care and study the mortise and tenon joints, then mark them to make the later assembling convenient. Dismantle every part in a reverse order of its assembling originally. During the dismantling process, maintain the integrity of the objects, try the best to avoid damage to the lacquer layer, especially preserving the original condition objects, or cause any new damages to the old furniture.

d) Depainting and dewaxing (except for preserving the original condition) : different measures are taken for removing the original paint and wax after dismantling.

e) Cleaning the dusts: the dust on the surface of the furniture should be cleaned .Organic solvent is forbidden to prevent the objects from fading colors. Cleaning is to maintain the original quality,

# 修复工艺流程

## 1. 修复前准备工作

①记录。首先对文物家具木器照相、进行文字描述，以记录下器物破损部位及破损程度。

②编制修复计划。根据家具的破损程度、编制完整的修复计划，制定相关的技术标准、填写工艺流程表。

③拆解。仔细观察器物结构，在器物的不同部位标上序号（方便以后安装），按照与原器物组装时相反的顺序把器物拆开。拆卸时要注意保持器物的完整性，尽量避免破坏漆膜（对于保旧工件要特别注意）或形成新的损伤。

④脱漆脱蜡（保旧工件除外）。器物拆御后，可采用不同方法去除旧器物的漆膜蜡膜。

⑤去脏保旧。清除表面浮灰，不得使用任何有机溶剂，以防变色。保持文物的原有特质，对清灰过程中散落的部件进行位置记录和

the position of the falling off pieces of the objects must be recorded and saved in safe place for later conservation.

The dismantled parts of the furniture should be sunk in warm water to remove the glue, dust and dirt remained at the mortise and tenon. The adhesion and durance would be greatly affected if old glue stays at its old place.

**Woodworking**
Wood needs to be selected, matched and processed.

i) Wood Selection
Examine the quality of wood, levels and the locations of deterioration of the objects that needs to be conserved. Then select the traditional material of the same quality and color, use the wood with similar quality and slightly lighter in color if the material with the same appearance cannot be found.

拆解
Disassembling

归缝
Gap Re-attaching

加固
Consolidation

刨边
Edge Trimming with Plane

修边
Edge Trimming

边缝清灰
Gap Cleaning

木刮刀去脏
Cleaning with Wood Scraper

气吹除尘
Dust Cleaning

清缝
Gap Cleaning

细节清灰
Detail Cleaning

保存，待修缮过程中恢复。

器物部件要用热水浸泡，以便把榫卯等处的胶、泥等污迹清洗掉，如果清洗不干净，修复后的牢固程度和修复效果将受到严重影响。

### 2.木工

工件在木工车间内一般要经过选料、配料、加工等程序。

（1）选料

观察器物本身的木质、损坏的程度及部位，选用同质、同色木材的老料，特殊情况时，要尽量使用木材花纹、肌量相近，颜色稍浅的材料。

（2）配料

根据加工要求和器物各部位的受力情况，进行合理配料，要求无裂，无疤节、无腐烂，保证器物的强度和外观一致性。

ii) Wood Matching

According to the stress of each part of the objects, the compensated wood should be match with the original wood in the matter of strength and appearance. And the wood must be free of crack, burl and rot.

iii) Wood Cutting

a) Line drawing. Before sawing the round log, the log should be marked with Chinese carpenter's line marker, which determines the outturn percentage of the log.

b) Wood sawing. The log would be sawed into board by two people with a saw. Generally, the outer layer would be skin. The inner section under skin would be sawed into panels with beautiful wood grain. And heartwood would be sawed into thick boards and later processed into legs and doorjambs.

iv) Wood Processing

The selected wood would be processed mortise

（3）制材

① 圆木弹线。用专用"墨斗"在圆木上弹线，这是锯解板材的依据，决定着用材是否合理和出材率的高低。

② 圆木锯板材。两人用手工大锯将圆木锯成板材，按线锯料，一般圆木靠最外出一膘皮，二膘出薄板，花纹美观，芯材出厚板，进一步变为腿枨等。

（4）加工

根据器物零部件的质量及技术要求，对选好的材料进行加工，比如加工各种形式开榫卯结构、打孔、钻眼、型面曲面加工等。工序可分为：画、锯、刨、打面、净光、画线、开榫凿卯、攒面刹肩、组装站活。

"京作"硬木家具常用的榫卯结构有龙凤榫、攒边打槽、三碰肩、穿肖挂榫、插肩榫、抄手榫、圆包圆、霸王枨、走马肖等数十种，以及由此演变而来的上百种结点方式。

（5）试装

将各部位修整后的零件进行初装，要保持原来外观各部件以及框架结构的严密、合理。

3.雕刻

雕刻往往是器物的画龙点睛之处，题材以祥兽瑞草和有吉祥喻意的组合图案。雕刻工序可分为以下几方面。

①画。将经过木工加工后需要雕刻的部件，按要求画出图案实样，经审查合格后贴在器物部件上。

②搜。将需要透雕的部件用搜弓锯搜出空透之处。

③凿。用凿子将器物部件按图案要求凿出图案胚型。

④铲。用刻刀将图案胚型铲出细致图形，转回木工，组装在器物有关部位。

木工和雕刻可以交叉进行操作。中国古家具木器中常用的传统雕刻手法有线雕、浮雕、透雕等，

and tenon structure, drilling and decorative surface trimming. The steps would follow line drawing, sawing, planning, polishing, and joining together. The joinery of Beijing-style furniture could be divided into dozens of joinery such as tongue-and-groove joint, inserted shoulder joint, hook-and-plug tenon joint, and further evolving into hundreds of connecting pattern between components.

v) Trail Assembling

Assemble the compensated part after the wood is trimmed. The structure and the component should stay tight.

**Carving**

Carving is considered to be the essence of an object. The themes of the carving decoration are mainly the combination of auspicious animals, plants and other auspicious patterns. The process of the carving technique is:

a) Pattern drafting: draw the pattern and attach the pattern to the processed wood after the pattern is approved.

b) Saw: saw the hollow section of the openwork carving pattern with bow-shaped saw.

c) Chisel: carve the shape of the pattern with chisel.

d) Shovel: shovel the pattern with carving tools. Then continue the woodworking process and assemble the components of the object.

The process of woodworking and carving could be mixed. Traditional Chinese carving may include line engraving, relief carving and openwork carving and so on. The steps of carving are pattern drafting, wood grain matching, chiseling and slicking.

**Polishing**

Polishing, one of the most crucial process, includes wood surface polishing and lacquer surface polishing. Usually, same amount of time would be used on polishing as wood processing, so that the

补后雕刻
Carving after Compensation

补配件
Loss Compensation

打胚
Shape Carving

雕刻
Carving

修花
Detail Carving

在操作上分图稿设计、配花、打坯、修光等工序。

### 4.打磨

打磨分为木材表面修整和漆面修整，是修复中最重要的工序之一。

一般情况下，打磨时间与加工时间基本相当，这样才能保证配件光洁度、手感等基本与器物相同。

### 5.髹饰

传统家具木器的髹饰主要包括制漆或制蜡、选料、刮灰、配色、髹漆髹蜡等。通常要经过打磨、着色、打磨、揩漆揩蜡、复漆复蜡、擦蜡等工序。一般情况下，一件硬木家具在头道漆或蜡后，要再上四到五道面漆或蜡，上两次色，揩漆、蜡和复漆、蜡一共需要八至十道，由木质的好坏而定，木质好的上面漆蜡和复漆蜡的道数可适当减少。

smoothness of the compensated part could match the original surface.

**Surface Coating and Painting**

Surface decoration of traditional furniture may include lacquer or wax making, plaster application, color matching, lacquer and wax application, etc. And the surface coating and painting need the procedures such like polishing, coloring, re-polishing, lacquer or wax application and wax polishing with dry cloth. Generally, after the first layer of lacquer and wax, the furniture would be painted with lacquer or waxed for another four to five layers, colored twice, and then apply eight to ten layers of lacquer or wax. Layer of the wax and lacquer is determined by the quality of wood, thus the better the wood, the less layer of wax or lacquer would be applied.

**Assembling**

The assembling of the furniture involves the assembling of metalwork, accessories of the objects and the final process of assembling of all

对页：修复工艺
Restoration Technique

### 6. 装配

修复后家具木器的装配主要包括金属件的安装、器物部件装配，以及最后的总装配等工序。总装配时要根据器物原有结构进行合理组装，做到榫卯结构严密，边框平直、无胶痕，表面光滑，四脚平衡。

### 7. 干磨硬亮

木质零件表面的修整，过去传统的打磨多用磨石，打磨前先用刀修整，然后用开水浇一下木胎，使木胎的毛孔显露，这样能够打磨的光滑。现在可按照先粗后细的原则进行打磨。

磨。分石磨、锉草磨两种，根据不同的木料和器物的不同部位分别使用。

石磨。将天然的磨刀石制成长方或方形平石块，用布将温水抹在器物平面和部件上，用平面磨石往返磨擦，将器物平面和部件平面处，磨平、磨细，达到光细平整。

锉草磨。锉草又称木则草，用沸水浸软，可弯曲成各种形状。内用一支扁形竹制挺棍，外裹锉草，沾水往返磨擦家具的线条和雕刻的花纹图案，花纹图案要磨细逼真，根底部要干净利落。

磨活有"凿一、铲二、磨三工"的说法，就是说如果雕凿用一个工，铲活就要用两个工，磨活则需要三个工。

家具木器漆膜蜡膜的打磨，使用沙叶和木贼草。厚的桐油漆面需用喷灯烤化漆面后进行打磨，较薄的漆膜蜡膜可用刮刀进行处理。打磨的总体要求是不留死角、不留油污、色泽均匀一致。

### 8. 烫蜡

烫蜡技术最先被应用在青铜器表面，据容庚《商周彝器通考》中记载："乾嘉以前出土之器，磨砻光泽，外敷以蜡"，可以让青铜器历经千年而不腐，具有很好的保护作用。另据宫中内务府造办处的《活计档》，其中多处记载了宫廷备料常有黄蜡、白蜡，黄蜡即现在常用的蜂蜡，白蜡即川蜡。后来这种技艺被有心的匠人加以利用，应用

the components. All the components need to be assembled on the basis of their original structure. During the process, the joinery should be tight, no remains of glue should be observed, surface should be smooth and the feet should stay balance.

**Dry-grinding Polishing**

When a piece of furniture is completed, each side of it is trimmed with knife first; then the wood appeared when hot water poured onto the surface. After these steps, the wood could be polished smooth enough. This is the traditional method. Now, it is polished step by step.

Polishing: stone polishing and dry-grass polishing, either is to be applied according to the materials and parts of the furniture.

Stone polishing: natural stone was cut into flat rectangular or square in shape; moist the wood surface with wet cloth, and then grind it with stone repeatedly till it becomes smooth and flat.

Dry-grass polishing: Cuo-cao is a type of dry grass, which could be folded into different shapes after sunk into hot water. Wrap grass around a piece of bamboo into a brush, dip water and grind the wood carving repeatedly in order to clear ditches and smooth the surface.

There is a saying: "chisel one, shovel two and polish three", that is to say, one-hour work of chisel will require two-hour to shovel and three-hour to polish.

The lacquer layer surface of wood furniture polishing is done with tree leaf and equisetum. Thick tong-oil lacquer layer has to be polished when melting, but thin layer will be polished by knife. Every part and corner has to be polished smooth and clear, no oil or dirt allows to be remained.

**Waxing**

Waxing is firstly applied onto the surface of bronze objects, Rong Geng said: "the bronze objects excavated before Qianlong and Jiaqing periods in Qing dynasty are smooth and waxed outside surface" in his *Study of the Bronze Sacrificial Vessels of Shang*

在小件的根雕作品上,随着技术的逐渐成熟,进而发展应用到家具木器表面。

烫蜡技术是中国明式家具进行木材表面处理的一种装饰方法,不仅能很好地展现木材优美的纹理,而且在木材表面形成了一层保护膜,以防止外界环境对木材的不利作用。随着时间的流逝,这层保护膜由于不断地受到空气的氧化、人手的抚摸和抹布擦试等因素的影响,使家具的表面、棱角和边线等处出现了一种自然的、透亮的、温润如玉的表面形态,即"包浆",产生"天然去雕饰,清水出芙蓉"的美感。

烫蜡一般是用蜂蜡,如没有也可用做蜡烛的白蜡。小面积烫蜡时可一手持蜡块,一手用电烙铁或火烙铁融化腊块,滴或涂到木器表面,使其形成薄薄一层。如果大面积烫腊,这道工序也可把腊块放入金属容器加热融成腊液,再用毛刷或汤匙将腊液涂或浇到木器上,这样可大大提高效率。然后用烙铁熨烫蜡层,以热力使腊液渗入木纹,这道工序最重要,一是必须充分融化腊液,适当保持加热时间;二是不能太过,以不烫焦木纤维为宜。全部烫完后,用稍粗的干布擦净浮蜡,这时可用电吹风机稍稍加热表面,这样擦净浮蜡会容易一些。最后用净布全面擦光木器。

烫蜡具体步骤分为:点蜡、布蜡、烫蜡、起蜡、擦蜡、抖蜡。

(1)点蜡

烫蜡的前期工作,目的是为擦蜡涂匀做准备。这里要说明一点,我们用的蜡是蜂蜡、川蜡加松香合成的混合蜡。在一般民间制造、修复家具中,为防止蜡凝固,都会加入一定比例的煤油稀释。但是由于文物修护的特殊性,防止煤油这种有机化学成分物质对木质本身和木理的伤害,因此所用的蜂蜡没有加入煤油,为防止蜡凝固,就需一次做少量的蜡,并在烫蜡的整个过程中要定时不断用高温风枪给蜡加温以防凝固,因此相比民间烫蜡要费时费力得多。点蜡就是用蜡刷(三排或者四排的鬃刷),均匀地向文物家

*and Zhou Dynasties*. The waxed bronze objects can be preserved so well from getting rotten after thousands of years. And *Work File of the Workshop of Imperial Household Department* recorded that there were beeswax, paraffin wax stored in the Imperial Palace. Waxing technique then was applied to the tree root caving objects by the craftsmen, and then even adopted to furniture surface as the technology developed gradually and skillfully.

Waxing is a kind of decoration method used for dealing with the furniture surface in Ming Dynasty. It cannot only present the beautiful grin of the wood, but also protect wood surface, so as to avoid the damage from environment. As time going on, the wax layer become natural, transparent and mild, after oxidized in the air, hand touching and clothing cleaning for long, thus the wood natural resin comes out to be considered as "wrapped slurry".

Beeswax is usually used for waxing, or sometimes paraffin wax. While waxing a small area, one hand holds a hot iron and melts the wax chunk in the other hand, drops of wax onto the wood surface to form a wax layer. But, when it is a big area, the wax chunk will be put into a metal container and heated to melt; apply the melted wax with brush or pour it directly onto wood surface; then, melt the wax layer with hot iron, so wax goes into the wood grin. This procedure is quite important, one must ensure the time and temperature to heat and melt the wax both are correct, and wood fiber shouldn't be burnt. After waxing, remove the surplus wax with a dry cloth, heat waxing layer slightly with hair drier which makes this easier. Polishing with clean cloth will finally make furniture appear to be smooth and bright.

Waxing procedure: wax-dripping, wax-spreading, wax-melting, wax-removing, wax-cloth-polishing, wax-brush-polishing.

i) Wax-dripping

This is the preparatory step for wax-spreading.

具表面的各部分依次点蜡。

（2）布蜡

就是把上步所点好的蜡，在均匀地加热过程中，均匀地铺开。注意：此步骤中一定要用纯棉布，切记勿用较硬的布料。

（3）烫蜡

对于文物家具木器，烫蜡的过程中温度要低，用蜡要少。过程是热风枪（故宫里不允许用明火）在一定的匀速抖动中将蜡化开，使其慢慢深入到材质里面去。要求：利用热风枪烫蜡过程当中一定要抖动，古旧文物家具的修复当中根据不同部位选择使用热风枪或者电炉丝。热风枪的优点是热量集中，灵活性强，易操作；其缺点是温度过高，不适合在镂雕或者材质疏散的家具表面过多使用。多数适用于面积较大的座面、踏面，腿部表面也可以使用。在镂雕部分更适合用散状放热的电炉丝。但不管是哪种工具都要求掌握距离、适当抖动来控制热量以免破坏家具表面。

（4）起蜡

首先使用特制的工具，工具传统是用牛角制成，但是在制造特殊形状的蜡铲上可以使用同材质木料制成的蜡铲（俗称蜡起子）。对于普通平面制成平面蜡铲：头部磨平便可；但对于有镂雕的部位就要制成锥形、凹形。比如在有线条部位起蜡，由于线条是凸出的，工具特意制成凹形，以方便起蜡。在起蜡过程中要注意做到一气呵成，目的是一次性铲净文物家具表面的浮蜡。也起到一定的压光作用。蜡起得越净，表面手感越细腻，圆润。

（5）擦蜡

在传统的手法上应该使用小粗布，纯棉布，如同搓澡样顺家具纹理进行戳擦，不但要有力还要有一定的速度，使摩擦过程产生一定的温度，使蜡不能定，这样可使没铲净的浮蜡去掉，手感直观上达到最佳。擦蜡完成后家具表面不能有粘手的感觉，这才能达到技术要求。

（6）抖蜡

也俗称干抖蜡。就是用猪鬃刷，要选择7~8

The wax is a mixture of beeswax, insect wax and rosin; usually adding paraffin can prevent the wax from concreting. However, paraffin is not applied because it will damage the wood grin to some extent and hot air should be heated enough to melt wax, this will slow down the work. Wax-dripping is so to drop wax evenly over the wood surface with coir brushes.

ii) Wax-spreading

This step will need one to spread the drips of wax melted over the surface, and one has to use soft cotton clothing.

iii) Wax-melting

The temperature will not be too high with small amount of wax on it when one is conserving a piece of cultural relic. One adopts hot air to melt wax slowly and evenly to make wax go into the wood grin. Hot air should not be very concentrated or high in case to damage the object. This method is usually applied to wax big areas of furniture surface such as seat, foot set, and legs. The open carving area will be more suitable for electric stove wires. Therefore, what kind the tools are proper temperature and skill are necessary for consideration.

iv) Wax-removing

First, a special tool, which is traditionally made of ox horn, is made according to the hardness of wood to remove wax. One end is made flat and thin to remove wax on flat surface; but different shapes of the tools are made for removing wax on open carving, they may be in the shape of cone, spill, and so on. Wax-removing must be done in one time evenly and clearly in order to feel fine and smooth.

v) Wax-cloth-polishing

Traditionally, a piece of coarse cotton clothing is applied to grind furniture surface. Certain speed and physical strength are required while grinding to cause warmth to clear up surplus wax, surface thus won't be sticky in the end.

厘米宽，12~15厘米长且硬度好的黑色猪鬃的板刷，但也不要使用野猪鬃，因为硬度过强。用刷抖动将各个部位擦净，包括缝隙里的存蜡。从文物家具的各个角度看，表面的光泽度是一致的，至此整个烫蜡工艺过程完毕。

松香混合蜡：松香的特点是具有隔水性和一定的粘合能力。在修复文物家具木器中，需要烫蜡 有些家具常年搁置，年代已久，材质非常酥松，在保护性修复时，就要适当补充水分，这点要求很高的技术性。当木头完全丢失时，可以说就几乎成"碳"了。不再有任何力量。因此要补充水分，然后使用大约不超过8%的混合蜡进行保护。同时对没有打开部位也是一种再加固过程。制作时，火候与混合比例有非常严格的要求，且具有很高的技术性。

### 9.擦蜡打光

烫蜡后的家具木器，仍需要用洁净白布折叠成硬布捲，在器物的表层用力抹擦，使之生热，将蜡溶渗到木丝棕眼内，全部涂平，冷却后能凝固在

点蜡
Wax Dripping

烫蜡
Wax-speading

起蜡
Wax-melting

抛光
Wax-removing

抖蜡
Wax-cloth-polishing

vi) Wax-brush-polishing
This step is to use bristle brush not wild boar bristle brush, usually 7 to 8 centimeters wide, 12 to 15 centimeters long, to grind all parts of surface, including the ditches to clear surplus wax.

The furniture looks quite even in color, and flat on surface after all the steps finished.

Mixed wax with rosin: rosin is able to keep water away from moisture and adhere. Therefore, the old furniture needs to be moister when conserve it, and wood structure becomes loose due to its long history. But high technique is required for moisture, because the heating temperature and proportion are hard to control that 8% or less mixed wax is needed for conservation. At the same time, the dismantled parts will be consolidated.

### Rub-on wax for polishing

Furniture has to grind with clean white cloths folded up into a roll after waxing as the last step, the grinding makes furniture surface warm enough

修缮前
Before Restoration

修缮后
After Restoration